MAHSEER FISH BIONOMICS AND POPULATION

Barrage Impact on Fish Biology

MAHSEER FISH BIONOMICS AND POPULATION

Barrage Impact on Fish Biology

by

Dr. DAVENDRA S. MALIK

Dr. KAMAL SINGH NEGI

2007

DAYA PUBLISHING HOUSE

Delhi - 110 035

ISBN 81-7035-368-8

Published by : **Daya Publishing House**
1123/74, Deva Ram Park
Tri Nagar, Delhi - 110 035
Phone: 27383999
Fax: (011) 23244987
e-mail : dayabooks@vsnl.com
website : www.dayabooks.com

Showroom : 4760-61/23, Ansari Road, Darya Ganj,
New Delhi - 110 002
Phone: 23245578, 23244987

Laser Typesetting : **Classic Computer Services**
Delhi - 110 035

Printed at : **Chawla Offset Printers**
Delhi - 110 052

PRINTED IN INDIA

Preface

Nature has endowed Garhwal region with wealth of flora, fauna and large area of water resources. The fluvial ecosystem of this region embodies the snowfed rivers, spring-fed tributaries and thousands of rivulets. The foothill region forms a unique transition zone as reflected in the biodiversity of the aquatic ecosystem. The Ganga river system is the largest and the most important riverine fisheries resource of Uttaranchal and Uttar Pradesh states. First time, Hamilton gave a description of 269 species of endemic fishes from the river Ganga in his pioneering work entitled "**Fishes of Ganges**".

Mahseer is one of the most important game fish of India and regarded as a sacred fish by the Hindus called it '**Mahisa**' meaning the fish per excellence. In hill-streams, the present trend of decline of mahseer's population due to indiscriminate fishing of brood and juvenile fish and the adverse effect of river valley projects. The existing dams or barrages across the Ganga river are posing threat to food-feeding habit, growth spawning biology and fish migration. In the present book, the authors have tried to put all observations on the record of mahseer biology affected by barrage on Ganga river at Hardwar-Rishikesh region and suggested rehabilitation and conservation measures of mahseer fishery in term of habitat restoration, *in-situ* stock enhancement and proper planning management of hydroelectric projects on Ganga river system. At last, we are sure, the book would be useful to all concerned and help in further development of mahseer fishery in upland streams.

D.S. Malik

K.S. Negi

Contents

Chapter 1
Introduction

The Indian mountains since time immemorial had remained place of worship and abode for the naturalists and spiritualists. Of late these mountains have become a place of recreation providing an outlet for indulging in both outdoor and indoor activities. The ever increasing population pressure on the mountain region adversely affected the fragile ecosystem. The important of saving the mountain environment and protecting the adverse biosphere was realized when International Mountain Society came into existence. Sensitive in their make up, tectonically, meteorologically and biologically they are most susceptible to slightest disturbance and the consequence can be catastrophic. High altitude zone especially Himalayas are highly sensitive with their very merge tolerance where steep slope, gravity coupled with climatic stress make their recovery impossible.

Geological history of the Indian sub-continent shows that the region now occupied by the Himalayas and the Indo-Gangetic plains south of them once formed the substratum of the Tethys sea.

The Himalayas which form an arc-shaped wall on the northern frontiers of India are about 2500 km in length and 150–450 km in width. This chain of mountains which lies parallel to each other can conveniently be divided into:

1. Greater Himalaya containing the highest peaks in India. The Nanga Parbat (8,126 m above msl) and Nanda Devi (7,817 m above msl).

2. Lesser Himalaya, which includes spurs and highly dissected uplands.
3. Shiwalik Himalaya, which constitute the southern most section containing rocks and broken chain of mountains.

The Himalayas are one of the greatest physical feature of the world. In fact it is a part of a chain of mountains beginning in the west with the Pyrenese and the Alps in Europe and extending right upto the eastern frontiers of India and much beyond into Burma. The Himalayas spread along northern borders of India and have a great impact on the geo-physiography, ethnology, history and cultural heritage of India. In this mountain chain that has a profound influence on the climatological and environmental fronts of Indian sub-continents.

The Himalayas represent the youngest mountain range, having been uplifted about 60–70 million years ago. It is also geologically a very complex range, and its origin is not yet clearly understood. The geology of the Himalaya is also very complex and some parts of it are practically a terra incognita due to the ruggedness of the terrain and the hostile conditions of the climate.

The glacier resources in the Himalayas are vast. The glaciers have their usefulness as water resources. The Himalayan glacier ranging approximately in latitude from 27°N to 36°N and longitude 72°E to 96°E, consists in the greatest part of the cryosphere in the low-mid latitude, beside the Antarctica and the North Polar Region. Although, the Himalayan glacier follows the common properties of the cryosphere, it also takes on significantly the special characteristics as its low-mid latitude position, higher elevation and undergoing the influences of several atmospheric circulation system. There are 5218 glaciers occurring an area of 38221 sq km, which is 9.04 per cent of the total area of Himalayan zone.

Himalayan glaciers are very important as they are the major factors in controlling the climatic changes in Himalaya and the country. Himalaya can thus be classified into area of:

1. Dominant monsoon precipitation.
2. Equal to sub-euqal monsoon and winter precipitation, and
3. Dominant winter precipitation.

The region of Mount Everest, Nepal is an example of the first, the area in Ganga basin (equal), in the Beas basin and in parts of the Chenab basin (sub equal) are of the second type, and the regions of Laddakh, Spiti and parts of Tibet are of the third type. Glaciers are also a major source of our water resources as they are feeding nineteen major rivers of north India. Snow melt and glacier melt contribute significantly to the spring season flows in the Himalayan rivers.

The Uttaranchal carved out from thirteen northern districts of the Uttar Pradesh was constituted as 27th state of the Indian Republic on November 9, 2000. The state is spread over a geographical area of 53,204 square kilometres with more than 80 per cent in mountainous region in the central Himalayas. The altitudinal and geographical variation, mountain slopes, expansion of river valleys and vegetation covers results in varying climates in different parts of Uttaranchal State. It is hot an humid (tropical) in the river valleys below 600 msl. At higher elevations, the climate becomes subtropical upto 1200 msl. It has cool temperature upto 1800 msl and cold temperature upto 2500 msl and at still higher altitudes the climate is extremely cold. It receives the highest rainfall in the range from 245 to 260 cm and records low temperature in majority of the regions. But the state is inadequately developed, less accessible thinly populated, nearly urbanized and physically isolated. The state has two main distinct division *viz.*, Garhwal and Kumaon.

Geographically, Garhwal has almost central position in the long Himalayan sweep, it is the most important part of lesser Himalaya, which lies between the latitudes 29° 26′ to 31° 28′ N and longtitude 77° 49′ to 80° 06′ E. It has an area of about 30,090 sq km, spreading 220 km in east-west and 235 km in north-south. On the whole region constitutes about 7.01 per cent of the entire Indian Himalaya. The Tons river separates it from Himachal Pradesh in the west and the district boundaries of Nainital, Almora and Pithoragarh separate it from Kumaon division in the east, starting from the foothills in the south, the region extends upto snow-clad peaks, making the Indo-Tibetan boundary in the north.

The climate of the Garhwal Himalaya very much diversified due to the fact of variation in altitude and physical features. On the southern borders of Garhwal Himalaya 80 to 90 per cent of yearly precipitation falls as rain during the period July to September. While in case of passing towards north, a large portion of annual precipitation fall during winter in the form of snow. The winter precipitation in the higher ranges of Himalaya might be important in forming the glaciers, out of which the most important rivers originate. These rivers are the most important and common wealth of Uttaranchal state in Himalayan zone.

Nature has endowed Garhwal region with wealth of flora, fauna and large area of water resources. The aquatic resources are present in the form of rivers, serpentine rivulets, streams, beautiful lakes and reservoirs. Some of the most important river systems of northern India owe their origin to the glaciers of the Garhwal Himalaya. The fluvial ecosystem of this region embodies the snow-fed rivers and streams notably Alaknanda, Bhagirathi, Yamuna, Bhilangana, Jar Ganga, Asi Ganga, Mandakini, Nandakini, Pinder etc.; large number of spring-fed tributaries and thousands of rivulet, which finally pour into major rivers at various sites. Besides, Garhwal hills are also beautifully dotted with a number of natural ice-cold lakes and small ponds. The nature has gifted a varies and colourfull aquatic flora and fauna in all these water bodies.

In the Garhwal Himalayan region, many big and small hydroelectric power plants are established for power generation. These power plants are beneficial as well as harmful in many ways for the fragile environment of Himalayan region. Diminishing fossil fuel resources and looming water shortages are focusing alternation on the potential for multi-purpose usage of the national surface, water resources. As the government plans to undertake a large number of runoff the river schemes and storage dams in the near future, it becomes highly necessary that the environmental, social and economic impacts of such project are assessed and highlighted, so the inferences could be drawn for our on-going and future projects.

The existing hydroelectric projects including those under construction and proposed shall be coming up in Garhwal region in near future a total of twenty projects, thirteen are ones in the Ganga and remaining seven on the Yamuna stream. In case of the former, quite a number of projects are under construction including the controversial Tehri and Vishnuprayag projects on the Bhagirathi and Alaknanda main stream respectively. Although these proposed and commissioned projects, seven on the Bhagirathi and four on the Alaknanda shall certainly influence the prestine quality of the Ganga including the fish fauna, it will be projects in the different zones of the Ganga and its main tributaries that shall be crucial for the survival of the Garhwal Himalayan mahseer (Nautiyal and Singh, 1989).

In the mountainous zone of Shiwalik Himalayas a number of tributaries having their sources at high altitudes, join the river Alaknanda *viz.* Dhauli Ganga at Vishnuprayag, Birahi Ganga between Vishnuprayag and Nandprayag, Nandakini at Nandprayag, Pinder at Karnprayag and Mandakini at Rudraprayag, finally it joins river Bhagirathi at Devprayag. Ganga, the mighty Indian river originates from the snow-clad peaks of Himalayas, is the life line of millions of Indians. From its source to its entry into the Bay of Bengal, it travels a distance of around 2525 km. The river with its well knit tributaries drains the Ganga basin, which encompasses an area of more than a million square kilometers (1060,000 sq km) spread over four countries India, Nepal, Bangladesh and China (Chaphekar and Mhatre, 1983).

The Ganga has by far the largest river basin in India (15th in Asia and 29th in World) draining as much as 861404 sq km within the country, covering more than a quarter (26.2 per cent) of India's total geographical area. The basin has large surface water and ground water resources. The annual flow in the basin is 468.7 billion cubic meters, which accounts for 25.2 per cent of India's total water resources. Out of this only 142.6 billion cubic meters is consumed in the basin. Irrigation alone accounts for more than 94 per cent of total basin consumption, whereas domestic and industrial sector use only 2.9 and 2.8 per cent respectively. Due to the much higher intensity of rainfall in the Himalayan region, the streams which join the Ganga from north contribute more than 60 per cent of the water flowing in the Ganga basin. The peninsular streams contribute the remaining 40 per cent water in southern region.

The main stream of the river Ganga, the Bhagirathi originates from the ice cave of Gaumukh (33° 55′ N, 79° 07′ E) at the snout of Gangotri glacier in the Garhwal portion of western Himalaya, at an elevation of about 4100 meters. The Alaknanda, the sister stream of Bhagirathi rises beyond Mana pass, 8 km away from Badrinath (30° 44'29" N, 79° 29'41" E, 3123 m above msl) joins it at Devprayag. It is below this confluence, the united stream is known as Ganga river. After a run of some 267 km, it emerges into Indo-Gangetic plains at

Hardwar (elevations 288 m), where it swells into a mighty system, 750 m wide (approximately).

As a result of many developmental activities and frequently natural ecodisaster, soil erosion and sedimentation in the Ganga basin have become a common and serious ecological prepositions. The upland tributaries, Bhagirathi and Alaknanda of Ganga basin are the examples where the coldwater fish habitats, spawning and feeding grounds have been eroded to a great extent by frequent flash floods, land slides and blockade formations. Some important fishes of Ganga basin have shown a sharp depletion in their population.

India is fortunate is possess vast and varied fish germplasm resources distributed widely in vivid aquatic ecosystem. The aqua resources of the country include 2.02 million square kilometer area of Exclusive Economic Zone of surrounding seas, more than 29,000 kilometer length of rivers, about 1,13,000 kilometer of canals, about 1.75 million hectare of existing water-spread in the form of reservoirs, about one million hectare in the form of tanks and ponds and about 0.6 million hectare of stagnant, derelict, swampy water-spread areas (Jhingran, 1991).

About 2,200 finfish species have been recorded from different ecosystem of India while more than 20,000 species recorded world-wide. Out of these, about 400 species are commercially important, which include cultured, culturable and wild species. These fishes have acquired a wide variety of forms and habitats which have been reflected in their adaptation to live in markedly varying biotypes, ranging from cold torrential mountain streams to the dark abyssal depth of the seas. The approximate ecosystem wise distributions of fish germplasm resources of India are: coldwater (73) 3.32 per cent, warmwaters of plain (544) 24.73 per cent, brackishwater (143) 6.50 per cent and marinewater (1440) 65.45 per cent (Annon, 1992–93; Das, 1994).

The principal endemic and exotic fish occurring in natural upland waters of the Himalaya are about 225 species. The mahseers, which belong to one of the groups among Indian carps (Cyprinidae), are clad with largest scales recorded in fresh water fishes. The group comes in spectrum of colours ranging from deep, burnt copper, through gold, silver and black. The amateur and professional anglers sport fisherman are well familiar with the mahseer group of the fishes. The commonly occurring mahseer species found in India are:

1. *Tor tor*
2. *Tor putitora*
3. *Tor mosal*
4. *Tor khudree*
5. *Tor musullah*
6. *Tor progeneius*
7. *Acrossocheilus hexagonolepis*

In past two decades, construction of dam or barrages across on many rivers for hydroelectric power generation, irrigation and navigation purposes have interfered with the riverine environment, inevitably affecting the fishery resources of the concerned rivers. The existing dam or barrages are posing threat to food and feeding habit, spawning biology and fish migration. They are altering the magnitude, chronology, quality of stream flow below dams and effecting fish stock. Drastic changes in fish habitat are taking place in the tail waters of these dams. Changes in physico-chemical and biotic parameters have been affected these migratory threatened fish population of mahseer. Many of the investigations pertain principally to basic work without considering the impact of dams and barrages, they act as a physical barrier in between aquatic ecology and mainly affected the population dynamics of the hill stream fishes.

Tor tor and *Tor putitora* species of mahseer commonly found in the Garhwal region of Uttaranchal Himalaya. But the population of sport fish mahseer has been continue to be decimated by the large scale destruction of their natural habitat and are categorised as threatened fishes in India (NCA, 1976). The biological studies also give original data on habit and habitat, sexual dimorphism, sexual maturity, catchable size, spawning period, spawning sites, normal fluctuation in year classes, population size distribution, fish production from catch data and experiments on natural breeding. Some other minor aspects of fish ecology and biology for proper management and fish production are also given. All these studies are essential to obtain a rational fishery of a particular species in a water body. This year to year knowledge on mahseer and its biological aspects, will help to evaluate the condition of stock in Ganga river. The present scientific study regarding the impact of barrage on the biology of mahseer specially population dynamics, feeding habits, changes the spawning and migratory behaviour, fish habitat distoration, and suggestive conservation measures would be provide a baseline to improve the present status of mahseer population in Ganga river at Garhwal region.

Chapter 2

Historical Resume

Mahseer (*Tor tor, Tor putitora*) are the most important game and food fishes of India, provides the main fishery in upland waters of the Shiwalik Himalaya and lower reaches of Deccan plateau. Somesvara mentions certain cold water fishes, notably Mahasila (Mahasirsha or Mahasiras, Barbus (*Tor*) mussullah Sykes and *B. Tor khudree* Sykes) which refers to the famous sporting fish of Indian Mahseer (Hora, 1951). Further, the Sutras and Smriti, two components of the Vedas mention of Mahasalka in Manu Smriti (III, 268). This is related to fishes possessing large scales exclusively including several species of the famous upland sporting fishes collectively going under the name of Mahseer. The commonest species Golden or Himalayan mahseer is described in Manu Smriti as the privileged fish relished by Saints dwelling in forests and by the Brahmins especially for propitiating the souls of the deceased ancestors for an endless period. Yajnavalkya Smriti (X, 258–261) was also in agreement with the above statement. Vishnu Smriti (LXXX, 14) regarded mahseer as a fish that should always be accepted (Hora, 1953).

Day (1876–1879) made reference to various species of mahseers occurring in India, Burma and Ceylon. In the early part of this century, all carps with big scales and barbels resembling in other morphological characters were grouped as mahseer. Perhaps, the detail work on the species of mahseer was dealt by Hora (1939–44) in a series of articles under the title of "*Game Fishes of India*". Sen and Jayaram (1982) and Talwar and Jhingran (1991) were identified six valid species under the genus *Tor* and one species under the, genus *Acrossocheilus.* The valid species of the genus *Tor* are the *Tor* mahseer, *Tor tor* (Hamilton), the *putitor* or the golden or the Himalayan mahseer, *Tor putitora* (Hamilton); The mussulah mahseer, *Tor mussulah* (Sykes); The yellow or khudree mahseer, *Tor khudree* (Sykes); The copper or mosal mahseer, *Tor mosal* (Hamilton) and the Boker of Assamese or the Katli of Nepalese, *Tor progeneius* (McClelland). Under the genus *Acrossocheilus,* the Chocolate mahseer, *Acrossocheilus hexagonolepis* (McClelland) was identified. Among the seven valid species of Mahseer, *Tor tor* and *Tor putitora* are the most common species found all along the Himalayan belt from Kashmir to Assam. Besides these *Tor tor* also occurs in rivers Narmada and Tapti in Madhya Pradesh and Chambal river in Rajasthan.

The study on the biology of mahseer, *Tor tor* and *Tor putitora* reported by Thomas (1879) in his historical account entitled *"Rod in India"*. This gives a complete account on the biology and sporting qualities of mahseer species.

Amongst the six valid species, golden mahseer (*Tor putitora*) is the largest in size attaining the length up to 270 cm and 170 lbs in weight (Mishra, 1962and Sen and Jayaram, 1982). Hamilton (1822) described first time golden mahseer under the name Cyprinus (*Cyprinus putitora*) and McClelland (1839) reported golden mahseer as *Barbus puritora.* Finally, Hora (1939) gave the systematic position of golden mahseer and recognised *Tor* as a subgenus under the genus Barbus.

The mahseers, which belong to one of the groups among Indian carps (Cyprinidae) are clad in the largest shining scales recorded in freshwater fishes. Therefore these are also referred as large scaled carps in India. Among the mahseers, the putitor mahseer or the golden mahseer is one of the widely distributed species of genus *Tor* extending from Hindukush-Kabul-Kohistan in the north-west to Sadiya in north-east. Besides this, the *Tor* species is also reported from Afghanistan, Bangladesh, Myanmar, Nepal, Pakistan, Sri Lanka and even Thailand and Ningpo in China (McClelland, 1844). The Scientific description of the mahseer fishes described by Hamilton (1822) in his famous book *"Fishes of the Ganges"*. Some general observations on mahseer were made by Walker (1888), McDonald (1936), Hora (1939, 1940), Smith (1947), Lal (1962, 1963), Desai (1970, 1973), Chaturvedi (1976), Das and Pathani (1978),

Jhingran and Sehgal (1978), Pathani (1981, 1982), Nautiyal and Lal (1985), Sharma (1987), Jhingran (1991), Sehgal (1991) nad Gupta and Khan (1992) in lentic and lotic waters.

Most of the mahseer species were recognized as endangered or threatened and general decline in their catches were reported in almost all natural waters of the Himalayas. The impact of dams, weirs, barrages etc. on the catches of mahseer *Tor putitora* wer studied by Khan (1940) and Rai (1947) in Punjab waters. Jhingran and Sehgal (1978) studied the fast decreasing trend in catches of golden mahseer in Punjab. Das and Pathani (1978) gave a brief note on the declining trend of mahseer in rivers of Kumaon Himalayas.

Hora (1921) reported some new and rare species of fishes from the eastern Himalayas. Fishes of the Eastern Doon, United province reported by Hora and Mukerjee (1936). Fish fauna of different river system were reported by Badola and Pant (1973), Badola and Singh (1977), Mahajan (1961), Menon (1962 and 1992), Lal and Chaterjee (1962), Sharma and Rajput (1982), Singh *et al.* (1983), Khanna and Badola (1990), Om Kumar *et al.* (1990), Joshi *et al.* (1993) and Malik *et al.* (1998).

Day (1871) reported *Tor putitora* in the rivers of Himalayas, Kashmir, Sikkim and Assam. David (1955) reported the mahseer availability in the Mahanadi river near the Huma sanctuary (South India). Hora and Mukerjee (1936) and Lal (1963) reported the occurrence of mahseers, *Tor tor* and *Tor putitora* in Song and Suswa river system at Doon valley. The problem of Kumaon mahseer observed by Pathani (1977) and Das and Pathani (1978) in lentic water of Kumaon Himalaya. Mahseer present in Beas, Sutluj and Yamuna river system reported by Joshi (1980, 1984 and 1988). Jayaram (1981) contributed wide distribution of *Tor tor* in the foot hills of Himalaya, Madhya Pradesh, Bihar, Assam, North Bengal, Pakistan and Bangladesh. Natural history and status of Himalayan mahseer in Alaknanda and Nayar river in Garhwal region recorded by Nautiyal and Lal (1982, 1984 and 1988) and Nautiyal (1984 and 1990). Dasgupta (1982 and 1991) reported the availability of mahseer in Brahmaputra river system in North Eastern region. Sharma (1987) also recorded some aspects of mahseer in Bhagirathi river system. Availability of mahseer *Tor tor* seed in river Narmada at Hoshangabad (M.P.) reported by Langer *et al.* (1992). Shrestha (1986 and 1994) reported the presence of *Tor putitora* in Himalayan water of Nepal. Present status of *Tor tor* and *Tor putitora* reported by Malik *et al.* (1998) and described diversity pattern and conservation of endangered fish species in upland waters.

Data on fisheries, fish catch and population size of any particular species are also essential for fishery management. In the present study, fish catch composition data have been made to evaluate the condition of mahseer fishery in river Ganga at Rishikesh. There are some contributions on population size and fish catch composition data for freshwater fishes in India and abroad. Shetter and Hizzard (1939) studied species composition by age groups and stability in fish population. Ricker (1946) described production and utilization of fish population. Carlander (1953) highlighted and Lake (1957) observed trout population and habitats in New South Wales: the main observation on fish population of clear lake. Krishnamurthy *et al.* (1964) observed on experimental fishing with gill nets in the Tungabhadra reservoir. Karamchandani and Pisholkar (1967) reviewed fish and fisheries of Tapti river. Further Sehgal (1971) described fisheries survey of Himachal Pradesh. Shah and Raizada (1977) highlighted the observation on creel census and angling pressure on river Bear. Mitra *et al.* (1977) studied Chilka lake fisheries. Payer and Scalet (1978) studied population and production estimates of *Pimephales promelas*. Power (1978) studied on fish population structure in arctic lakes. Joshi *et al.* (1978) observed on fishery resources of the hill streams of Jammu province. Das and Pathani (1978b) studied affect of lake ecology on the productivity of mahseer *Tor tor* and *Tor putitora* in Kumaon lakes. Pathani (1984) studied population biology of mahseer *Tor tor* and *Tor putitora* (Hamilton). Srivastava *et al.* (1985) reported inland fish marketing in India. Pathani (1988) studied biology and ecology of mahseer *Tor tor* and *Tor putitora* (Hamilton). Langer *et al.* (1992) collected the fry at different centres in river Narmada. Gupta and Khan (1994) reported mahseer fisheries in India. Desai (1992) reported fisheries of Sukta reservoir in the context of mahseer fish and Nautiyal *et al.* (1998) studied role of mahseer in magnitude of commercial fishery in Garhwal hills.

Biometric studies are useful for the identification of a fish species and for detecting variations in the fish population. Biometry reflects the proportionate growth of different body parts and the influence of environmental factors in a particular habitat. Several workers have studied in biometric characters in different water bodies of India and abroad. Schaefer (1948) studied morphometric characters of yellow fin tuna (*Neothunnus macropterus*). Schaefer and Walford (1950) reported biometric comparison between yellow fin tunas (*Neothunnus macropterus*). Pillay (1957) reported morphometric study of the population Hilsa, *Hilsa ilisha* (Ham.). Marr (1955) studied on the use of morphometric data in systematic and relative growth. Royce (1963) reported morphometric study of yellow fin tuna, *Thunnus albacore* (Bonnaterre). Dwivedi and Menezes (1974) studied a note on the morphometry and ecology of *Brachiurus orientalis* (Block and Schneider). Alam (1977) described the morphometric studies of a Indian Mackerel *Rastrellinger kanagurta* (Cuvier, 1817). Berg (1979) studied on external morphology of the pink salmon (*Onchorhynchus gorbuscha*). Coad and Power (1974) studied on meristic variation in three spine stickleback *Gasterosteus aculeates.* Libosvarsky and Ruban (1985) studied on morphometric deviations in Roach from heated waters. Dubey and Dubey (1986) examined biometric studies of Indian mahseer, *Tor tor* (Ham.). Lindsey (1988) reported controlling meristic variation in fishes. Dasgupta (1989a) studied biometry of the mahseer *Acrossocheilus hexagonolepis* (McClelland). Dasgupta (1989b and 1991a) studied biometry of the mahseer, *Tor tor* (Ham.) and *Tor putitora* (Ham.) collected from Garo hills. Sharma and Tondon (1990) reported the morphometric and meristic characters of *Cyprinus carpio* (Linn.). Wiencaszed (1992) studied biometric characteristic of Baltic eel-pout, *Zoarces viviparus* (L.). Menon (1992) studied on taxonomy of mahseer fishes of the genus *Tor* (Gray). Pandey and Nautiyal (1997) examined the statistical evaluation of some meristic and.morphometric characters of *Schizothorax richardsonii* (Gray) and *Schizothorax plagiostomus* (Heckel) and Singh and Singh (2000) studied on morphometric relationship of *Labeo calbasu* (Ham.) collected from the Ganga river.

The importance of age and growth studies in fishes is well established for the fishery biologist, which not only gives present, past and future population growth pattern of the fishery, but also the annual environmental fluctuation in which the fish exist. The determination of age of fishes by means of scales, operculum, vertebra and otoliths are in extensive usage.

Such studies have made by many workers in the past in India and abroad. The main workers are Lee Rosa (1920) reviewed on age and growth determination of some fishes by scale. Backman (1942) studied annuals formation on the scales of certain Michigan game fishes. Jhingran (1957, 1968 and 1971) studied on probable age of *Cirrhinus mrigala, Catla catla* and *Septinna phasa*. Frost and Kipling (1959) recorded age and growth of pike *Esox lucius* (L.) from scale and operculum bones. Chugunova (1959) studied age and growth in fish. Das (1959) elucidated first time that scales ring (annuli) can be used safely as age indicator even in tropical fresh water fishes and recorded age by scale method in twenty four fishes of India. Nickolsky (1963) studied scale, otoliths and age of fishes. Quasim and Bhatt (1966) examined age of *Ophiocephalus punctatus*. Bilton and Jenkinson (1968) studied on scale and otoliths of *Onchorhynchus nerka* and *O. chum* and Hile (1970) studied body scale relation and calculation of growth in fishes.

Ricker (1971) reviewed the methods for age and growth studies in details. Moriarty (1973) observed age of the eel by otoliths. Lear and Pitt (1975) estimated age of *Reinhardtius hippoglosoides* by otoliths. Jhingran (1977) studied on age of *Gudusia chapra* by scales. Das and Pathani (1978) studied age and growth of Kumaon mahseer, *Tor putitora* (Ham.). Black and Black (1978) reported the use of operculum bones in study of age and growth of *Labeo senegalensis*. Pathani (1977, 1979 and 1981) studied age and growth of *Tor tor* and *Tor putitora* by scale and otoliths. Johal and Tondon (1981, 1983, 1985, 1987, 1989 and 1992) studied age, growth and length-weight relationship in *Tor putitora* (Ham.), *Catla catla, Labeo rohita* and *Cirrhinus mrigala* in lentic and lotic waters. Erickson (1983) analyzed age determination of *Manitoban walleyes* using otoliths, dorsal spine and scales. Tondon and Johal (1983 and 1996) studied age and growth of *Tor putitora* (Ham.) and other Indian fresh water fish. Gjosaeter *et al.* (1984) studied ageing of tropical fish by growth rings in the otoliths. Radtke (1984) observed scanning electron microscope evidence for yearly growth zones in giant blufin tuna, *Thunnus thynnus*.Nautiyal (1985 and 1990) observed length-weight relationship, relative condition factor, growth rate and age composition of Garhwal Himalayan mahseer in Alaknanda river. Tondon *et al.* (1989 and 1993) studied age and growth of *Labeo dero* and *Tor putitora* (Ham.) collected from Govindsagar reservoir. Dasgupta (1991b) examined length-weight relationship and condition factor (k-value) of *Tor putitora* (Hamilton) collected from Garo hills, Meghalaya. Sandhu *et al.* (1994) reported growth studies of *Tor putitora* (Ham.) in Govindsagar reservoir. Badapanda (1996) studied biology of *Tor mosal mahanadicus* (David). Khillare and Skore (1998) observed age by otolith in eel-pout, *Zoarces viviparus* collected from Hel Peninsula, Poland in relation to reproduction. Sinha *et al.* (1999) studied length-weight relationship and condition factor of a thermal spring loach, *Lepidocephalichthyes guntea* (Ham.). Mohan and Saraswat (2000) studied length-weight relationship of *Cyprinus carpio* from Jodhpur and Bali and Sharma (2002) studied length-weight relationship and relative condition factor of *Tor putitora* (Hamilton) in river Beas.

Many research contributions are available on food and feeding habit of fishes. The major ones are Thomas (1873) described the Rod in India; Allen (1935) studied the food and migration of the pearch, *Perca fluviatilis* in windermere. Mukerjee and Ghosh (1945) studied food and feeding habits in major carp. Mukerjee and Sengupta (1946) observed food and correlation with body weight and length of gut in *Cirrhina reba*. Al-Hussaini (1947) studied the feeding habits of some telostean fishes. Hartle (1948) studied food and feeding relationship in a community of freshwater fishes. Hynes (1959) reported the food of freshwater Sticklebacks with a review of methods for food estimation. Pillay (1951 and 1953) gave critic of food study methods and feeding habit of *Mugil cephalus*. Alikunhi (1952) reviewed feeding on the young carps. Mishra (1953) worked on the gut contents of *Labeo rohita, Cirrhinus mrigala* and *Catla catla*. Das and Moitra (1955, 1956, 1958 and 1965) reported food and feeding habits of twenty four fresh water fishes of Uttar Pradesh. Barrington (1957) reported the alimentary canal and digestion of fishes. Menon and Chacko (1958) described food and feeding habits of some fresh water fishes of Madras. Khanna (1961) contributed on the foods of some major carps of North India. Ivlev (1961) studied experimental ecology of the feeding in fishes.

Some data on food and feeding habits were also given by Lal and Chatterjee (1962) in Eastern Doon fishes. David (1963) studied feeding behaviour of *Pungasius pungasius*. Nikolskii (1963) studied the ecology of fishes. George (1963 and 1965) reported food and feeding of Delhi fishes. Khanna and Pant (1964) reported notes on food of some freshwater fishes of Kumaon. Qayyum and Qasim (1964) studied on the biology of three freshwater fishes of Aligarh. Jhingran (1968 and 1972) studied food and feeding of *Catla catla* and *Gudusia chapra* in Ganga river. Ishiwata (1969 a,b) recorded ecological studies on the feeding of fishes. Desai (1970) reportedfeeding of *Tor tor* coolected from Narmada river. Kapoor *et al.* (1975) studied the alimentary canal and digestion in teleost fishes. Das and Pathani (1978) contributed on the biology of Kumaon Mahseer, *Tor putitora.* Pathani and Joshi (1980) reported food and feeding habits of fingerlings of the *Tor tor* and *Tor putitora*. Hyatt (1979) examined feeding strategy of fishes. Badola and Singh (1980) studied food and feeding habits of fishes of the genera *Tor, Punitius* and *Barilius.* Sampath (1984) recorded preliminary report on the effects of feeding frequency in *Channa striatus*. Nautiyal (1990) studied the feeding ecology of *Tor putitora* in some hill streams at Garhwal. Pisholkar and Karamchandani (1981) recorded the food and feeding habits of *Tor tor* from Govindgarh lake at M.P. Ribble and Smith (1983) recorded relative intestine length and feeding ecology in freshwater fishes. Nautiyal and Lal (1984b and 1985a) studied food and feeding habits of mahseer. Sharma (1983, 1986 and 1987) reported food and feeding habits of *Chrossocheilus latius latius, Schizothorax plagiostomus* (Heckel) and mahseer, *Tor tor* (Hamilton) in Garhwal Himalaya. Dasgupta (1988, 1990a,b and 1993) studied the food and feeding habits of copper mahseer (*Acrossocheilus hexagonolepis*), *Tor tor* and *Tor putitora* (Hamilton). Badola *et al.* (1991) reported food and feeding habits of some hill-stream fishes of Garhwal Himalaya. Desai (1992) studied food of *Tor tor* (Hamilton) collected from four localities. Smith *et al.* (1993) observed daily and seasonal patterns in the feeding behavior of Atlantic salmon *Salmo salar* (L.). Malik (1994) studied feeding behavior in fingerlings of mahseer *Tor putitora* (Hamilton) and Boujard (1995) studied on diel rhythms of feeding activity in European catfish, *Silurus glanis.*

Studies on the sexual dimorphism and sex ratio of the existing endemic fishes are important not only in breeding biology but also helpful in assessment of population dynamics of fishes in the freshwater system. The predominance of particular sex in different periods of the year have made by many workers from India and abroad. Some recent and past works consulted by the author are Hickling and Rutenberg (1936) examined ovary as an indicator of spawning period. Khan (1939) studied sex organ of mahseer (Barbus *tor*). Smith (1947) reported sex of mahseer and Bokar. Mishra (1950) studied sexual dimorphism in *Labeo boggut* (Syke). George (1959) studied for distinguished the sex of oil sardine, *Sardinella longiceps*. Chaudhari (1959) noted external characters distinguished sex of common Indian carp. David (1963) reported sexual dimorphism in *Mystus gulio* (Ham.). Mraz (1964) studied sex ratio and maturity in the white fish. Das (1967) observed gonads and fecundity of food fishes. Rai (1967) studied on cyclical changes in the ovary of *Tor* (Barbus) *tor* (Hamilton). Spanishlab (1968) described the sex ratio in brown trout, *Salmo trutta fario*. Swamp (1968) studied sexual dimorphism in *Xenetodon cancilla*. Rao (1969) studied sexual dimorphism in fishes of family Copolidae. Saunders and Power (1970) examined sex ratio and fecundity in America smelt, *Osmerus mordax* (Mitchill). Swarup and Singh (1971) studied sexual dimorphism in the puffer fish, *Chelonodon cutcutia*. Swarup and Srivastava (1972) and Swamp *et al.* (1972) reported sexual dimorphism in *Giant gaurani; Colisa fasciatus* (Anabantidae) and spiny eel, *Mastacembelus panacalus*. Swamp and Swaroop (1975) and Swamp and Srivastava (1975) reported sexual dimorphism in catfish *Mystus vittatus* (Bloch) and *Pseudeutropius artherenoides* (Bloch). Pathani (1978) studied secondary sexual characters in Kumaon mahseer *Tor tor* (Ham.) and *Tor putitora* (Ham.) and James and Baragi (1980) examined ovary as an indicator of frequency of spawning in fishes.

Natural breeding and spawning behavior of fishes are an important aspects for fishery research as well as for proper conservation management of fisheries, because it is directly related to the evaluation of fish production. Several investigators have studied the problem of natural breeding and spawning ecology in some fishes of commercial importance. Nevill (1915) described the breeding habits of mahseer, Barbus *tor*. Wood (1930) examined pre spawning and spawning movements in Scottish herring shoals. Khan (1942); Hora (1945) and Hussian (1945) studied the spawning behavior of carps in Punjab. Das and Dasgupta (1945) also studied on breeding of the principal carps. Smith (1947) studied the natural spawning of trouts. Ganapati and Alikunhi (1950) reported the spawning of carps in the river Tungbhadra. Simpson (1959) observed spawning in the place of the north sea. David (1959) studied spawning grounds of the Gangetic major carps. Qasim and Qayyum (1961) examined spawning frequency and breeding seasons of freshwater fishes. Lal (1963) reported distribution and migration of *Tor putitora*. Bhatnagar (1964) examined spawning and fecundity of fishes in Bhakra reservoir. Kulkarni (1971) studied spawning habits of Deccan mahseer *Tor khudree*. Quasim (1973) studied maturation and spawning of marine teleost. Sobhana and Nair (1974) observed the maturation and spawning of *Puntius sarana subanastus*. Sundar and Joshi (1976) studied spawning of *Tor putitora*. Chaturvedi (1976) observed spawning biology of mahseer *Tor tor*. Das and Pathani (1978) studied the breeding and spawning biology of *Tor putitora*. Pathani (1977) observed spawning ecology of Kumaon mahseer *Tor tor* and Pathani (1980) studied spawning ecology of *Tor tor* and *Tor putitora* in Bhimtal lake. Badola and Singh (1980) studied on spawning of some cold water fishes in Garhwal hill streams. Nautiyal and Lal (1982) examined fecundity of the Garhwal Himalayan mahseer *Tor putitora*. Nautiyal and Lal (1984) reported migratory behaviour of the Himalayan mahseer in the rivers of Garhwal. Vasudevappa and James (1980) observed maturity and spawning of marine catfish, *Tachysurus dussumieire*. Thakre and Bapat (1981) reviewed maturity and spawning of *Rasbora daniconius*. Singh *et al.* (1985) observed spawning patterns and environmental regulation on the spawning in hillstream fishes. Shrestha (1986) reported spawning ecology and behaviour of *Tor putitora* in Himalayan water of Nepal. Dasgupta (1987) studied maturity and spawning in *Acrossocheilus hexagonolepis*. Nautiyal and Lal (1988) reported destruction of spawning grounds of mahseer in Himalayan rivers. Dobriyal and Singh (1989) studied spawning ecology of *Puntius chilinoides*. Desai (1993) examined reproduction biology and spawning ecology of *Tor tor*. De *et al.* (1994) studied impact of farakka barrage on the spawning of *Hilsa tenualosa ilisha* (Hamilton). Shrestha (1994) studied migration of golden mahseer in Himalayan waters of Nepal. Joshi and Sundar (1995) described breeding and culture of Snow trout, *Schizothorax richardsonii* (Gray) in Kumaon hills. Gaur and Pathani (1996) studied spawning ecology of *Barilius vagra* (Hamilton) and Malik *et al.* (2002) studied spawning periodicity of mahseer *Tor putitora* in Ganga river at foothills.

The fish population is shrinking in different habitats due to cumulative effect of various stresses. These stresses need to be removed for conservation of the fish population. The study on fish conservation have been made by many workers in the past in lentic and lotic waters of India. Dubey (1986) have stressed with important observation about conservation of mahseer in Narmada basin. Kulkarni (1986) described the conservation methods of mahseer resources in Lonawala. Das (1988) reported conservation of genetic fish resources in India. Menon (1988) described conservation of ichthyofauna and conservation of inland capture fisheries in India. Shrestha (1988) reported ranching of mahseer in Nepal. Nautiyal (1989) described problems and prospects of mahseer conservations in Garhwal regions. Jhinaran (1990) stated restoration of fisheries in Ganga river system. Kulkarni and Ogale (1991) described conservation of mighty mahseer through seed enhancement process. Das and Lakra (1992) reported conservation of fish genetic diversity in India. Kulkarni (1992) described role of aquaculture on the conservation of mahseer. Soni (1992) described conservation of mahseer. Nath *et al.* (1994) reported conservation of sport fishes in Doon valley region. Das (1994) described conservation of threatened fishes in India. Das (1996) described conservation of Himalayan mahseer. Malik *et al.* (1998) reported conservation of endangered fish species in upland waters. Bhaumik and Saha (1998) described role of mass awareness and public participation in fish conservation. Joshi (1994) studied fish conservation in Ramganga river. Kapoor *et al.* (1998) described ichthyodiversity of India as status and conservation and Badola and Badola (1999) described recommendations to protect fish fauna in Garhwal Himalaya.

Chapter 3

Physio-Geographical Feature of Study Area

Rishikesh is a pilgrim township of district Dehradun (Uttaranchal) situated in the foothills of Shiwalik Himalaya on latitude 30° 06′ N and longitude 78° 19′ E. The height of the Rishikesh is 356 meters above mean sea level, consisting area approximately 11.20 sq km. It is surrounded on three sides by Himalayan mountain ranges. Rishikesh is very well known from medieval time as celestial abode as a great pilgrimage centre of India, attracting a large number of pilgrims from India as well as foreigners. It is believed that Bharat, brother of Lord Rama subjected himself to serve penance here. A temple known as Bharat temple was later constructed at the bank of river Ganga and the town of Rishikesh sprang up around the temple. It has since been the abode of saints and sages from where they sojourn before undertaking the arduous pilgrimage that lay ahead on the distant Himalayan peaks.

The geographical feature of this Himalayan region consists of crystalineous and granites, gneisses and metamorphosed sediments, ranging in age from the pre-Cambrian to as late as the Miocene. Towards the north of this range occurs a pile of sedimentary rocks and south hills crystalline axis occur mixed zones of sedimentary rocks. Which are highly folded and faulted as metamorphics. The northern contact of these sedimentaries with the central crystalline a well marked tectonic feature, along with the crystallines are thought to have partially moved over the sedimentary zone. Similarly, the southern boundary of this zone is marked by another major tectonic feature, known as the main boundary fault. To the south of this fault occurs the autochthonous zone of tertiary rocks, comprising the molassic deposits of the Shiwalik group. The Shiwalik range are still highly wooded in many parts. They mainly comprise soft tertiary sediments sandstones, siltstones, shales and clayes are readily subject to mechanical disintegration. The ranges being in the front also face the full force of the monsoon current and have torrential rains. Hence, mass-wasting is very common in these areas and all along huge landslides or scars of old landslides, bringing in tonnes of material and posing a constant hazard finally to Ganga riverine system.

On the right bank of Ganga river, Rajaji National Park has spread over an area of 820.42 sq kms. Geographically, it is located between the latitude 29° 51′ N to 30° 15′ N and the longitude 77° 52′ E to 78° 22′ E. The park is 13 km in south-east of Dehradun, 9 km distant from Hardwar and 250 km distant from Delhi. Three sanctuaries in the Dehradun Shiwalik-Rajaji, Motichur and Chilla were amalgamated into a large protected area and named Rajaji National Park. The Shiwalik ranges of hills are formed of unconsolidated Himalayan and sub Himalayan debris, which is mainly in the form of alluvial gravels and coarse soil.

The climate of the Rajaji National Park is temperate and humid. The principal determinants of natural vegetation includes, climate, soil, relief and biotic factors. Forests constitute the most valuable natural wealth of the region. This zone is dominated by deciduous and sub-deciduous type of vegetation. Sal (*Shorea robusta*) is the most significant species of this zone, which found upto an altitude of above 4,290 feet. A few other species found in this zone include Bakali (*Anogeisus latifolia*), Sain (*Terminalia tomentosa*). Middle story comprises of Sandan (*Ougeinia oogenensis*), Duddhi (*Holarrhena antidysenterica*), Khair, Sisso, Semal, Kanju, Haldu, Rohini, Kusum and Bamboo etc. Ground cover comprises of herbs and grasses.

The river Ganga makes a steep descent with an average slope of 30.6 m km^{-1} in the first 70 km upto Nandprayag of its longitudinal profile. Afterwards its gradient is not so steep in the remaining hills with an average slope of 3.04 m km^{-1}. The river has formed deep valleys, sometimes narrow and sometimes broad in accordance with lithology. Being supported by large nowfields and glaciers, all the Himalayan tributaries of the Ganga river are characterized by well regulated flow and assured supply of snowmelt water throughout the year.

Ganga enters first time in plains in the Rishikesh. The flow of Ganga varies enormously seasonally. The winter season is the lean flow period. Increase in temperature in summer months, the flow of river is increase due to melting of ice from the higher altitudes. In the monsoon, the river flow comprises direct storm runoff. At this time, the flow values are several times higher than the rate in the lean flow period. After Rishikesh, the Ganga turns south-westwords of another 30 km in foothills and cuts across the Shiwalik range towards Hardwar to debounch on to the vast Indo-Gangetic plains. In this zone river is dammed by Pashulok barrage and, diverted into Chilla canal at 4 km downstream from main township of Rishikesh (Figures 1–3).

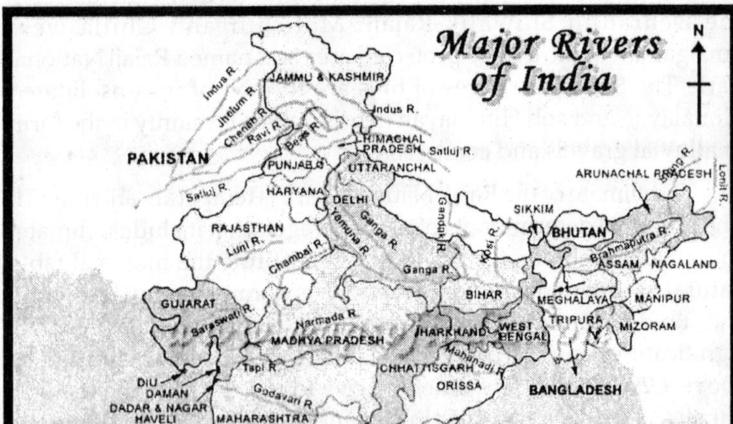

Figure 1: Riverine Resources of India

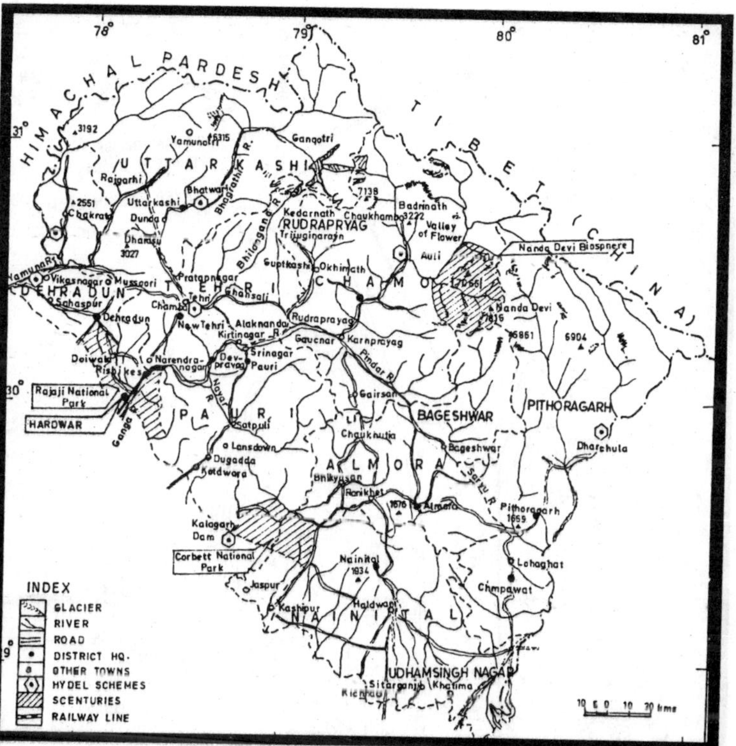

Figure 2: Riverine Resources of Ultaranchal State

The Pashulok barrage situated in right side of Delhi-Hardwar-Rishikesh National highway. The barrage was constructed across the Ganga canal at Kunnao just outside boundary of the Rajaji National Park. It is mainly established for Chilla Hydroelectric Power Generation Plant. This power plant was set up in 1970s on the Ganga canal, which flow parallel to the Ganga river and the capacity of power production is 144 mega watt. Chilla or Shakti canal is the water source for Chilla Hydroelectric Power Plant, the 14 km Kunnao-Chilla power canal was constructed on the last bank of the river Ganga. It is 22 meter wide with 45° sloped cement banks and the water is found throughout the year in this canal.namely, upstream,

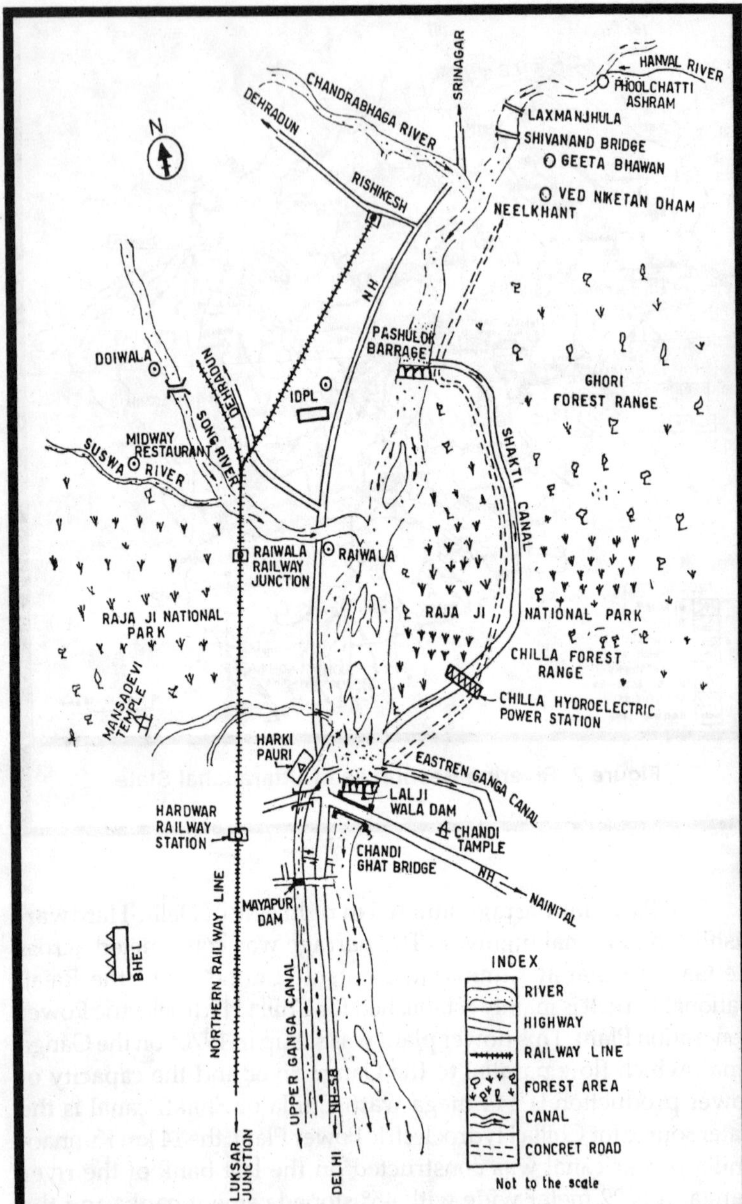

Figure 3: Map of Ganga River and its Tributaries of Study Zone with Showing Sampling Stations

The diverted river water again joins the main headstream of river Ganga after 21 km downstream at Bhimgoda in Hardwar region. In the leanflow season, the river in this 13 km zone have the only flow from river Suswa and Song, which joins the Ganga at Satyanarayana (upwards of Saptsarovar). At Hardwar 60 per cent of its water is diverted again into upper Ganga canal leaving little flow in the Ganga for next 80 km or more towards Uttar Pradesh.

Description About Sampling Stations

Site I

First sampling station is about 500 meter upwards from barrage. Due to dammed of main river, the Upstream form a reservoir which has large quantity of water found throughout the year. The substratum of this site is made by gravely covered by sand, silt and detritus (Plates 1 & 2).

Site II

Second sampling station is about 500 meter downwards from barrage. Due to diversion of water for Chilla canal, the quantity of water remains very low or sometime partially dried the Downstream. At this site the substratum is stony, pebbley and gravely which was covered by sand (Plates 3 & 4).

Plate 1: Upstream of Ganga River of Pashulok Barrage at Rishikesh During Rainy Season

Plate 2: Upstream of Ganga River of Pashulok Barrage at Rishikesh During Winter Season

Site III

Third sampling station is selected near Saptsarovar in main Ganga river between Rishikesh and Hardwar. The quantity of water

Plate 3: Downstream of Ganga River of Pashulok Barrage at Rishikesh (Sampling Station 2) During Winter Season

Plate 4: Downstream of Ganga River of Pashulok Barrage at Rishikesh During Summer Season

is much more than Downstream due to join of Song and Suswa river into Ganga river at Satyanarayana (just above the sampling station) (Plate 5).

Plate 5: Saptsarovar site of Ganga River in Between Rishikesh and Hardwar (Sampling Station 3)

Chapter 4

Materials and Methods

The methods adopted in investigating various biological aspects are based on standards evolved in different research laboratories both in India and abroad. Whenever necessary, the methodology was modified to suit the local conditions. The scientific observations on the biology of mahseer in Ganga river, were carried out by dividing the Ganga river stretch into three sampling sites namely upstream, downstream and Saptsarovar in Rishikesh region of Shivalik Himalaya.

Meteorological Parameters

The meteorological data, *i.e.* maximum and minimum air and water temperature, relative humidity and rainfall have been collected from the State Irrigation Meteorological Research Laboratory, State Forest Department and Central Water Commission at Rishikesh, which had all the facility to record all meteorological observation on daily basis.

Fish Catch Composition

The mahseer fishes, *Tor tor* and *Tor putitora* were collected seasonally during March 2000 to February 2002 from different sampling sites of Ganga river. The fish catch samples were obtained from local fish market, fisherman and anglers. The use of cast net was common to land the mahseer (small size) while gill nets in addition to hooks were used to land the larger size of mahseer in Ganga river system. The fishes were taken to Fish Biology and

Conservation Laboratory, Department of Zoology and Environmental Science, Gurukul KangriUniversity Hardwar, in fresh condition for further investigation and analysis.

The fish samples were preserved in 5 per cent formaline solutions and they were identified according to Day (1878, 1879), Gunther *et al.* (1959), Menon (1962), Badola and Pant (1973), Jayaram (1987) and Sehgal (1992).

Biometric Analysis

In order to determine biometry, 18 morphometric and 7 meristic measurements were calculated during the study period. The measurements were made in relation to total length and head length on the basis of methods provide by Low McConnel (1971), Dwivedi and Menezes (1974) and Jayaram (1981) as considering the following parameters:

Total Length (TL)

Total length was measured from the tip of the snout to the tip of the caudal fin of the fish.

Standard Length (SL)

Standard length was measured from the tip of the snout to the base of the caudal fin of the fish.

Head Length (HL)

Head length was measured from the tip of the snout to the posterior bony extremity of the operculum of the fish.

Length of Snout (SnL)

Snout length was measured from tip of the snout to anterior end of the orbit.

Body Depth (BD)

The vertical measurement from a point in the body of the fish on its back where its height is greatest to a straight line to the ventral profile.

Eye Diameter (ED)

Eye diameter measured from margin of the body orbits of fish.

Inter Orbital Distance (IOD)

Inter orbital distance was measured along the dorsal surface between the eyes.

Length of Rostral Barbel (LRB)

Length of rostral barbel was measured from base to tip of the rostral barbel of fish.

Length of Maxillary Barbel (LMB)

Length of maxillary barbel was measured from base to tip of the maxillary barbel of fish.

Length of Caudal Fin (LCF)

Length of caudal fin was measured to the length of largest fin ray of caudal region of the fish.

Depth of Caudal Fin (DCF)

Distance of tip of Caudal fin upto base of caudal fin.

Length of Dorsal Fin (LDF)

The greatest distance between measured in a straight line between the anterior most and posterior most points junction with the body.

Length of Pectoral Fin (LPF)

The length of pectoral fin was measured by the length of its largest fin ray.

Length of Pelvic Fin (LPel.F)

The length of pelvic fin was measured by the length of largest fin ray of pelvic.

Length of Anal Fin (LAF)

The greatest distance in a straight line between the anterior most and posterior most point of the junction with body.

These measurements were subjected to correlation and regression analysis. The coefficient of correlation (r) and regression (b) were tested for significance. The data was then used to compute the regression equation for each other dependent variable to fit the straight line equation:

$$(y = a + bx)$$

where, y was the dependent variable 'a' the intercept 'b' the slope of the regression line and x is the independent variable.

Meristic Characters

Some Meristic characters were calculated as rays count of dorsal fin, pectoral fin, ventral fin, anal fin, caudal fin, lateral line scale counts and (girth scale counts) scale of transverse section were counts.

Age and Growth Determination

Age determination of mahseer fishes were analysed by scale and opercular bone method. Scale was also used for scanning electron microscopy in Instrumentation Centre of Indian Institute of Technology (IIT) at Roorkee.

Scale Method

The cycloid scales were collected from particular region of the fish body by the help of a pair of forceps. The key scales were taken from caudal peduncle and between lateral line and anal fins. Freshly pulled scales were washed in tap water and rubbed between thumb and index finger to remove the epithelial layer and the chromatophores. These scales were kept in 1 per cent KOH solutions for a few seconds, rubbed between fingers and cleaned in water to remove attached tissues. The cleaned scales were finally kept in 10 per cent formaline for two hours and then dried. These dried scales were preserved in separate envelops, which also contained the fish data such as total length, weight of fish and month and season of collection etc.

The methodology were applied for preparation of scales and study of age and growth have already been described by Lagler (1959), Chugunova (1963), Johal and Tondon (1985) and Tondon and Johal (1996).

In the present study back calculation of fish length by scale method was determined by applying the following formula (Dahl Lea, 1910):

$$L = as^2$$

$$L = a + cs^2$$

where,

L = Total length of fish;

s = Scale length;

a and c = are the constant.

Operculum Method

The entire operculum from each measured mahseer fishes were pulled out by strong forceps. The opercular bone was removed from the operculum of the some selected fish, which body scale were taken for age determination to compare and confirm the age of the fish. The opercular bone was boiled in hot water at 60°–70°C for 2–5 minutes to remove the adhering tissues and washed and dipped again in xylol, then finally sun dried and kept in the same envelop as the scale of the same fish for age determination.

The observations were calculated for counting the age of the mahseer fishes, the dried opercular bone was dipped in water and projected over a screen by micro-projector. The year marks were seen as opaque narrow ridge lying in between clear bands.

Condition Factor (K-factor)

The monthly and seasonal condition factor (K-factor) of fishes were determined by the formula:

$$K = W \times 10^5 / L^3$$

where,

K = Condition factor;

W = Weight of fish;

L = Total length

Food and Feeding Habit

The food and feeding habits of mahseer were analysed by gut analysis methods. The fresh fishes were dissected and the entire alimentary canal was removed from body cavity. The length and weight of alimentary canal and intestinal bulb were recorded. The measured intestinal bulb was then cut longitudinally by scissors and entire gut content was preserved in 5 per cent formaline solution for quantitative and qualitative analysis. Various scientific methods have been used described by Hynes (1950), Pillay (1951), Nikolsky (1963), Desai (1970) and Badola and Singh (1980).

Quantitative Analysis

The Sedgewick rafter method was used for quantitative study of fish food collected from the alimentary gut. Only the intestinal bulb content was considered as gut content for quantitative accuracy (since food estimation in small and large intestine is usually inaccurate). One millilitre of homogenized food from total gut content was taken in a Sedgewick rafter slide. The number of different items were multiplied by the dilution adopted at the beginning of each count. The sum of total points counted for the items of fish food were considered for percentage calculations.

Qualitative Analysis

The food volume of each fish was studied by the settling method. The entire gut content was kept in 15 ml graduated centrifuge tube or if large taken in 50 ml measuring cylinder. The settlings were done for 24 hours, after which the volume was read off and recorded for each fish in the fish catches of every season *i.e.* winter, summer and rainy. The food volume in intestinal bulb was recorded as full, three by fourth full, half full, one by fourth full or empty, and the guts were given points 20, 15, 10, 5 and 0 respectively as followed by Toor (1964), Pathani (1977) and Nautiyal and Lal (1982). The procedures of qualitative analysis were depends on eye estimation of distension of intestinal bulb and amount of food in gut.

Macroscopic and microscopic examination of the gut content were made to identify the food items. In spite of this, to ensure the qualitative analysis of the fish's diet, the intestinal as well as rectal portion were also examined.

Gastrosomatic Index (GSI) and Relative Length of Gut (RLG)

The gastrosomatic index (GSI) was estimated by the following standard formula (Bhatnagar and Karamchandani, 1970):

$$\text{GSI} = \frac{\text{Total Weight of Alimentary Canal of Fish with Food}}{\text{Total Fish Weight}}$$

The relative length of gut (RLG) was estimated by the usual method (Al Hussaini, 1949):

$$\text{RLG} = \frac{\text{Length of Entire Alimentary Canal}}{\text{Total Length of Fish in Fresh Condition}}$$

Sexual Dimorphism and Sex Ratio

Sexual Dimorphism

In the breeding seasons male and female of mahseers, *Tor tor* and *Tor putitora* were observed and recorded from Ganga river and its tributaries in Rishikesh-Hardwar region on the basis of many morphological and other characters as follows:

External Morphological Characters

Colour, size, fins and tubercles.

Other Characters

Body streak in breeding season.

Sex Ratio

In the breeding season, on the basis of total mahseer fish catch compositions the female and male specimens have differentiated by external morphological characters or pressing the abdominal belly to see the oozing condition in male specimens and releasing the mature and immature ova in female specimens. The confirmation test were made through dissecting the specimens to find out the position of gonads in breeding seasons.

Location of Natural Spawning Grounds

In order to locate the spawning grounds of mahseer, the eggs (fertilized) and fry were collected with the help of long handled dip plankton net from different spawning sites of Ganga river and its tributaries *i.e.* Hanval, Song and Suswa river at Rishikesh and Hardwar region of Garhwal Himalayas.

The physico-chemical and biological characteristics of the riverine water on spawning grounds were determined according to Welch (1948), Edmondson (1959), Needham and Needham (1966) and APHA (1995). The abiotic and biotic parameters give an idea of the optimum conditions required for successful spawning in natural conditions. Some physico-chemical parameters *viz.*, water temperature, water velocity, turbidity, dissolved oxygen, total hardness, pH, calcium, magnesium and chloride were analysed. The details of measurements of different physico-chemical parameters as follows:

Water Temperature

Water temperature was recorded with the help of (± 1°C) centigrade thermometer.

Water Velocity

Water current was recorded by float method.

Turbidity

For the turbidity measurement, the water samples were brought to the laboratory and turbidity was measured in NTU (Nephelometer Turbidity Unit) by using Nephelometer (made by Elico Pvt. Ltd., Hydrabad).

Apparatus and Reagent

Nephelometer (Turbidity meter)

It measured the scattered at the right angle of the path of incident light.

Samples Tubes

Stock Turbidity Suspension

1. Dissolved 1.0 gm of Hydrazine Sulphate $(NH_2)_2.H_2SO_4$ in distilled water to prepared 100 ml of solution.
2. Dissolved 10.0 gm Hexamethylene Tetramine $(CH_2)_6N_4$ in distilled water to prepare 100 ml of solution.

Mixed 5 ml of each of the above solutions (a and b) in a 100 ml volumetric flask and allow to stand for 24 hours at 25±3°C. Diluted it to 100 ml mark for prepare 400 NTU suspension.

Standard Turbidity Suspension

Prepared 40 NTU solution by diluting 10 ml of stock solution to 100 ml. This solution should be prepare every week.

Procedure

1. Set the instrument at 100 with the 40 NTU standard suspension. In this case, every division on the scale will be equal to 0.4 NTU turbidity.
2. Shake thoroughly the sample and kept it for sometime to eliminate the air bubbles.

3. Put the sample in the Nephelometer sample tube and find out the value on the scale.
4. If sample was suspected to have more turbidity (more than 40 NTU) dilute it and take the reading.

Calculation

Turbidity (NTU) = Nephelometer reading × 0.4 × dilution factor

Dissolved Oxygen

Dissolved oxygen of spawning grounds was measured by Winkler's Idometric method.

Reagents

Sodium Thiosulphate 0.025 N

Dissolved 24.82 gm of $Na_2S_2O_3 . 5H_2O$ in boiled and cooled distilled water and make up the volume to 1000 ml. Add 0.4 gm of borax or a pallet of NaOH as stabilizer. This is 0.1 N stock solution. Dilute it to 4 times with boiled and cooled distilled water to prepare 0.025 N solution.

Alkaline Potassium Iodide Solution

Dissolved 100 gm of KOH and 50 gm of KI in 200 ml of distilled water.

Manganous Sulphate Solution

Dissolved 100 gm of $MnSO_4.4H_2O$ in 200 ml of distilled water and filter.

Starch Solution

Dissolved 1.0 gm of starch in 100 ml of warm (80°C–90°C) distilled water and add a few drops of formaldehyde solution.

Sulphuric Acid

Concentrated H_2SO_4 (Sp. gr. 1.84).

Procedure

1. Filled the sample in a glass stoppered bottle (BOD bottle) of known volume carefully, avoiding any kind of bubbling and trapping of the air bubbles in the bottle after placing the stopper.

2. Poured 1 ml of each $MnSO_4$ and alkaline KI solutions, well below the surface from the walls. A precipitate will appear.
3. Place the stopper and shake the contents well by inverting the bottle repeatedly. Keep the bottle for some time to settled down the precipitate.
4. Add 1.2 ml of concentrat H_2SO_4 and shake to dissolved the precipitate.
5. Take 200 ml solution in a conical flask and add few drops of starch indicator solution and titrate with Sodium thiosulphate solution. At the end point, blue colour to transparent.

Calculation

$$\text{DO (mg/l)} = \frac{\text{ml of Sodium Thiosulphate used} \times N \times 8 \times 1000}{\text{Sample Volume Taken}}$$

Total Hardness

Hardness is the property of water, which prevents the lather formation with the soap and increase the boiling point of water. Total hardness of water analysed by EDTA titration method.

Reagents

EDTA Solution 0.01N

Dissolved 3.723 gm of disodium salt of EDTA in distilled water to prepare 1000 ml of solution. Store in Polyethylene or Pyrex bottle.

Buffer Solution

1. Dissolved 16.9 gm Ammonium chloride (NH_4Cl) in 143 ml of concentrated Ammonium hydroxide (NH_4OH).
2. Dissolved 1.179 gm of disodium salt of EDTA and 0.780 gm of $MgSO_4.7H_2O$ in 50 ml distilled water.

Mix both a and b solutions and dilute to 250 ml with distilled water.

Eriochrome Black T Indicator

Mixed 0.40 gm of Eriochrome black T with 100 gm NaCl and grind.

Procedure

1. Take 100 ml sample in a conical flask.
2. Add 1 ml of buffer solution for adjust pH 9–10.
3. Add 100–200 mg of Eriochrome black T indicator, the solution turns wine red.
4. Titrate the contents against EDTA solution. At the end point colour change from wine red to blue.

Calculation

$$\text{Hardness (mg/l)} = \frac{\text{ml of EDTA used} \times N \times 50 \times 1000}{\text{Sample Volume Taken}}$$

Calcium

Calcium hardness of waters of spawning site was analysed by EDTA method.

Reagents

EDTA Solution 0.01 N

Dissolved 3.723 gm of disodium salt of EDTA in distilled water to prepare 1 litre of solution.

Sodium Hydroxide 1 N

Dissolved 40 gm of NaOH in distilled water and dilute to 1000 ml.

Murexide Indicator

Mixed 0.2 gm of Ammonium purpurate with 100 gm of NaCl (A.R.) and grind.

Procedure

1. Take 100 ml sample in conical flask.
2. Add 2 ml of 6N NaOH solution for adjust pH 12–13 and add 100–200 mg of murexide indicator, pink colour develops.
3. Titrate against EDTA solution until the pink colour changes to purple.

Calculation

$$\text{Calcium (mg/l)} = \frac{\text{ml of EDTA used} \times N \times 40 \times 1000}{\text{Sample Volume Taken}}$$

Magnesium

Magnesium was analysed by EDTA method.

Reagents

EDTA Solution 0.01 N

Dissolved 3.723 gm of disodium salt of EDTA in distilled water to prepare 1000 ml of solution.

Buffer Solution

1. Dissolved 16.9 gm Ammonium chloride (NH_4Cl) in 143 ml of concentrated Ammonium hydroxide (NH_4OH).
2. Dissolved 16.9 gm of disodium salt of ETDA and 0.780 gm of $MgSO_4.7H_2O$ in 50 ml distilled water.

Mix both a and b solutions and dilute to 250 ml with distilled water.

Eriochrome Black T Indicator

Mix 0.40 gm of Eriochrome black T with 100 gm NaCl (A.R.) and grind.

Procedure

1. Find out the volume of EDTA used in calcium determination.
2. Also find out the volume of EDTA used in hardness (Ca^{++} + Mg^{++}) determination with same volume of the sample as taken in the calcium determination.

Calculation

$$\text{Magnesium (mg/l)} = \frac{x - y \times 24 \times N \times 1000}{\text{Sample Volume Taken}}$$

where,

x = EDTA used in hardness determination.

y = EDTA used in calcium determination for the same volume of the sample.

pH

pH is the negative $\log_{10}$ of the Hydrogen ion concentration in a solution. The pH value is expressed as the negative ion of $[H^+]$ concentration. pH of water sample was analysed by pH meter (portable) made by Hanna instruments.

Procedure

1. pH meter was calibrated with the help of buffer solution of 7.0, 4.0 and 9.2 pH.
2. Take the sample of water in a beaker.
3. Put the pH meter probe into water.
4. Switch on the pH meter and note the reading.

Chloride

Silver nitrate reacts with chloride to form very slightly soluble white precipitate of AgCl. At the end point when all chlorides get precipitated, free silver ions react with chromate to form silver chromate of reddish brown colour. Chloride was analysed by titration method.

Reagents

Silver Nitrate 0.02 N

Dissolved 3.400 gm of dried $AgNO_3$ (A.R.) in distilled water to make 1000 ml of solution and keep in a dark bottle.

Potassium Chromate 5 per cent

Dissolved 5.0 gm $K_2Cr_2O_7$ in 100 ml distilled water.

Procedure

1. Take 100 ml sample in an erlenmeyer flask and add 2 ml of $K_2Cr_2O_7$ solution.
2. Titrate the contents against 0.02 N $AgNO_3$ until a persistent red ting appears.

Calculation

$$\text{Chloride (mg/l)} = \frac{\text{ml of Silver Nitrate used} \times N \times 35.5 \times 1000}{\text{Sample Volume Taken}}$$

Biological Parameters

Analysis of Plankton (Phytoplankton and Zooplankton)

The plankton of the water in spawning grounds were collected by using conical hand plankton net. At the cone end, a specimen tube of 10 ml capacity was tied. The known amount of water was filtered and concentrated in the desired quantity, 0.1 ml of concentrated sample was taken and placed on a Sedgwick rafter counting cell. Plankton were counted by mirotransect method given by Lackey (1938) and Edmondson (1974).

$$n = \frac{(a \times 1000)\, c}{l}$$

where

n = Number of plankton per litre of water.

a = Average number of plankton in one small counting chamber of Sedgwick rafter counting cell.

c = Volume taken of plankton concentrated water.

l = Volume of total water filtered through plankton net.

Macrobenthic Density

The benthic biota of the river were sampled by Ekman's dredge size 30 × 30 × 30 cm at each spawning site besides this forcep were also use for collection of macrobenthos. The samples were sieved through a brass sieve of a mesh size of 0.5 × 0.5 mm. The invertebrates retained by this sieve were identified after Pennak (1953), Ward and Whipple (1959) and were counted numerically and computed per square meter as follows:

$$N = \frac{n}{ah}$$

where,

N = Number of macro-organisms in one square meter.

n = Number of macro-organisms per sampled area.

a = Number of Ekman's dredge in m^2.

h = Number of hauls.

Natural Breeding and Spawning Behaviour

The methods were applied to determine the natural breeding and spawning behaviour of mahseer (*Tor tor* and *Tor putitora*) were as follows:

Percentage of Occurrence of Mature Fishes

The percentage occurrence of mature fishes in the different months give an idea of the breeding season. The occurrence of the various maturity stages of fishes in different months were calculated by grouping the specimen of same maturity in each breeding months and calculating their percentage of matured female brood stocks in natural riverine system.

Occurrence of the Eggs and Larvae

Eggs and larval stages were observed only after spawning. The spawning periods were recorded on the basis of presence of spawns on spawning grounds.

Spawning Behaviour

During the breeding season in the early morning about 6:00 to 8:30 A.M. and evening 7:00 to 9:00 P.M. the earlier mating behaviour such as peduncle grasping, a trembling and chessing activities have closely observed on the different spawning grounds. The migratory behaviour in the early breeding season have also noted as the movement towards shallow waters on the shoreline of the Ganga river. The frequency of migratory movements of the male and female mahseer fishes on all selected spawning grounds have noticed. The confirmation test has followed by collecting the fertilized ova from the spawning grounds in breeding season, the diameter of the ova were calculated and the developmental process were carried out in the hatching tray put in the natural river system and calculated the time period of hatching, percentage of mortality, incubation period and cumulative survival rate.

Measurement for Conservation Practice of Mahseer Fishes

A weekly ground truth survey were carried on both the shoreline of Ganga river and its tributaries during the study period. The main Ganga river stream from upstream zone to Saptsarovar or upto the Bhimgoda barrage at Hardwar were divided into different zones for the conservation and management of the mahseer fishery. During the survey, some destructive fishing methods have noticed which were used by different group of society and fisherman to catch the fishes in day or night hours. Secondly, some natural breeding habitat were identified as well as restoration of few natural breeding habitat were also made for better spawning sites of the mahseer fishery. A large number of seeds as fingerling size were also collected from the Vyasghat in upper reaches of the Ganga river and stocked at Saptsarovar sampling stations as *in situ* conservation programme. Except these above programme, during the study period few number of mahseer conservation youth groups were created from the native locality of the adjoining villages of Ganga river.

Chapter 5
Observations

The study on the population of mahseer, *Tor tor* and *Tor putitora* and their biological aspects in the Ganga river was carried out from March 2000 to February 2002. The results of different parameters are briefly described as follows:

Meteorological Parameters

Rishikesh is situated in base of the foothills of Shiwalik Himalaya, climate of Rishikesh and its surrounding areas is relating colder than plains and can be categorized as sub-tropical. The minimum air temperature (12.39°C) was recorded in January during 2001–2002. From March to June, the air temperature has showed an increasing trend but sometime it has fluctuated and reaching to the maximum (29.61°C) in May 2001–2002. The average maximum air temperature was lowest 24.42°C in January 2001. The highest air temperature was recorded 39.62°C in May 2001. The values of average relative humidity recorded in the range of 39.00 per cent to 73.00 per cent and 38.50 per cent to 73.50 per cent in 2000–2002 respectively. In 2000–2001, the lowest relative humidity 39.00 per cent was recorded in April and highest 73.00 per cent in August month. In 2001–2002, the lowest value of relative humidity was recorded 38.50 per cent in June and highest value 73.50 per cent was recorded in August. The annual rainfall was recorded 1310.30 mm in 2000–2001 and the wettest month of the year was August with total rainfall calculated 518.80 mm. The total number of rainy days were 90 and maximum number of rainy days (22 days) during July month of 2000–2001. In 2001–2002, the annual rainfall observed 1465.10 mm, while the wettest month during the year was August with total rainfall amounting to 553.50 mm. The total number of rainy days was 93 in this year, while maximum numbers of rainy days (17 days) were June and August. (Tables 1, 2 and Figures 4a,b and 5a,b).

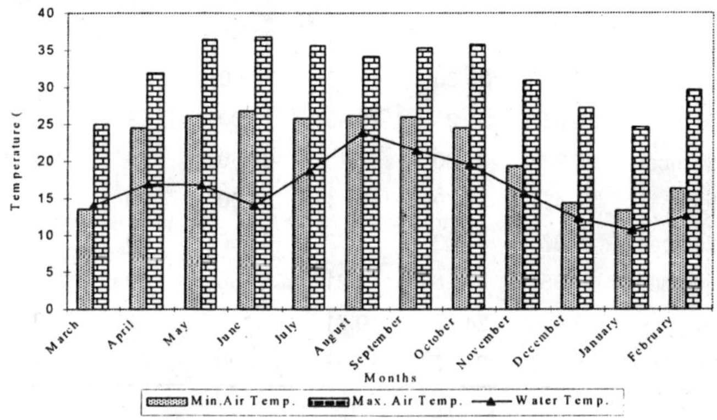

(a)

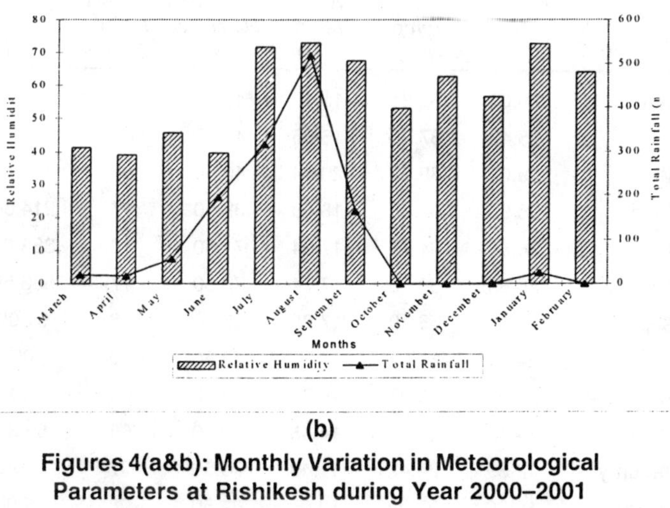

(b)

Figures 4(a&b): Monthly Variation in Meteorological Parameters at Rishikesh during Year 2000–2001

Table 1: Monthly Meteorological Data of Rishikesh during the Period Year 2000–2001

Month	Min. Air Temp. (°C)	Max. Air Temp. (°C)	Water Temp. (°C)	Relative Humidity (%)	Number of Rainy Days	Total Rainfall (mm)
March	13.60	25.06	14.05	41.00	04	20.30
April	24.54	32.00	16.92	39.00	03	16.40
May	26.20	36.53	16.77	45.50	11	55.80
June	26.82	36.77	13.97	39.50	16	195.00
July	25.81	35.60	18.68	71.50	22	316.10
August	26.05	34.21	23.95	73.00	19	518.80
September	25.93	35.30	21.40	67.50	13	164.50
October	24.45	35.79	19.44	53.00	00	00
November	19.37	31.03	15.62	62.50	00	00
December	14.35	27.26	12.21	56.50	00	00
January	13.41	24.67	10.71	72.50	02	23.40
February	16.23	29.73	12.57	64.00	00	00

Table 2: Monthly Meteorological Data of Rishikesh during the Period Year 2001–2002

Month	Min. Air Temp. (°C)	Max. Air Temp. (°C)	Water Temp. (°C)	Relative Humidity (%)	Number of Rainy Days	Total Rainfall (mm)
March	20.07	33.81	15.10	48.00	04	18.50
April	25.43	37.28	16.95	41.50	02	11.00
May	29.61	39.62	20.24	40.50	08	51.50
June	26.95	36.38	18.88	38.50	17	214.30
July	26.83	34.71	17.94	71.50	16	383.60
August	25.85	30.10	17.74	73.50	17	553.50
September	26.15	36.79	17.85	53.00	08	61.00
October	24.14	35.03	16.32	49.50	00	00
November	20.05	30.97	15.77	60.00	00	00
December	14.15	27.73	14.45	72.50	06	65.80
January	12.39	24.42	11.49	70.50	07	37.90
February	15.30	27.09	12.09	61.50	08	68.00

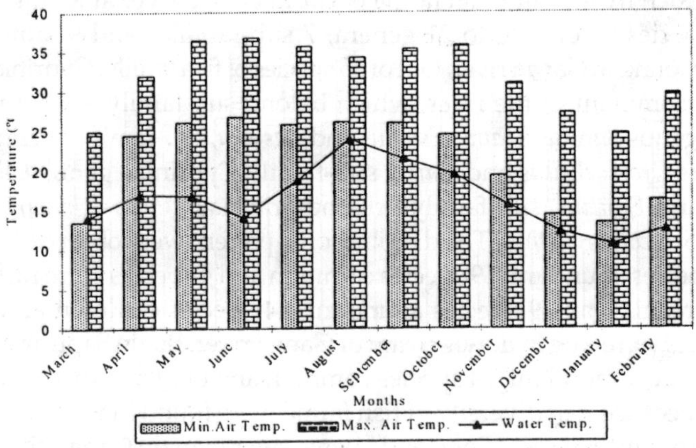

(a)

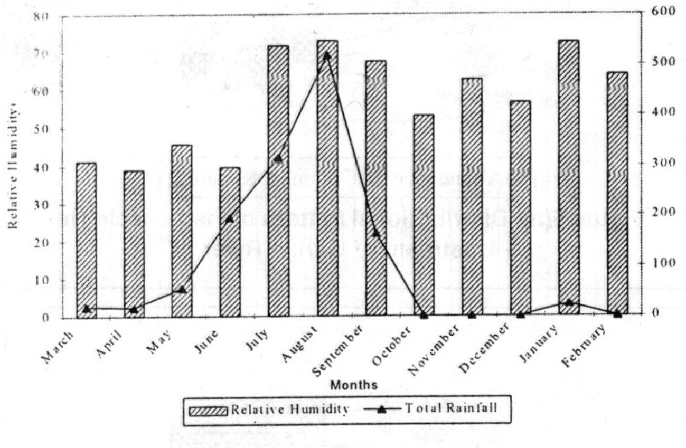

(b)

Figures 5(a&b): Monthly Variation in Meteorological Parameters at Rishikesh during Year 2001–2002

Ichthyodiversity

Rich fresh water fish fauna occur in Ganga river at Rishikesh. 35 species belonging to 21 genera, 7 sub-families and 9 families were found in Ganga river. Out of 35 species of fish family, Cyprinidae was dominant in the river, which belong sub-family Rasborinae and genus *Barilius, Danio, Esomus* and *Rasbora,* sub-family Garrinae, genus *Crossocheilus* and *Garra,* sub-family Cypriniae genus *Labeo, Tor* and *Puntius,* sub-family Schizothoracinae genus *Schizothorax* and *Schizothorachthys.* The distributional pattern were observed that 12 species abundant, 19 species common and 4 species were rare in upstream, while 9 species abundant, 14 species common and 12 species were rare in downstream of Ganga river. But in Saptsarovar, 20 species abundant, 14 species common and only one species was rare because rich diversity of fish fauna were found in Saptsarovar due to confluences of Song and Suswa river in main Ganga river at Satyanarayana in Hardwar region (Tables 3, 4 and Figs. 6a,b & c).

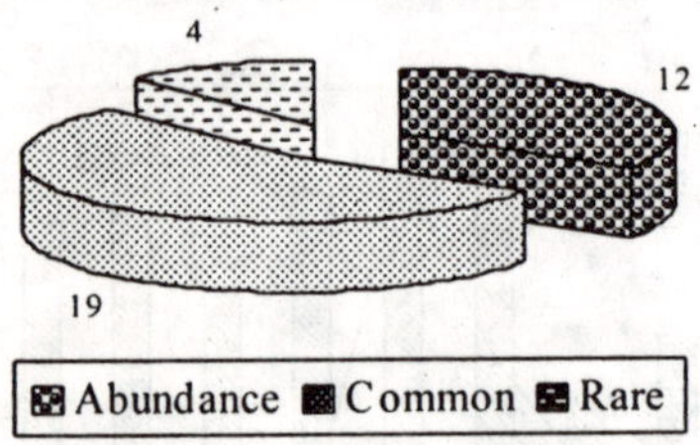

Figure 6(a): Distributional Pattern of Fish Species in Upstream of Ganga River

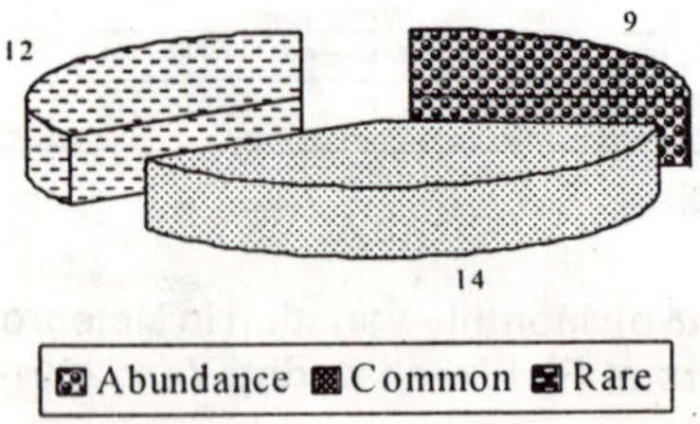

Figure 6(b): Distributional Pattern of Fish Species in Downstream of Ganga River

Table 3: List of Ichthyofauna in Ganga River

Present Scientific Name	*Name According to Day*	*Name According to Hamilton*	*Local Name*
Barilius bendelisis (Ham.)	*Barilius bendelisis*	*Cyprinus (Barilius) bendelisis*	*Chal*
Barilius bola (Ham.)	*Barilius bola*	*Cyprinus (Barilius) bola*	*Bhola*
Barilius vagra (Ham.)	*Barilius vagra*	*Cyprinus (Barilius) vagra*	*Persee*
Danio (Danio) deverio (Ham.)	*Danio devario*	*Cyprinus (Danio) devario*	*Patukari*
Danio (Brachydanio) rerio (Ham.)	*Danio rerio*	*Cyprinus (Danio) rerio*	*Patukari*
Esomus danricus (Ham.)	*Nuria danricas*	*Cyprinus (Danio) danrica*	*Dend*
Rasbora daniconius (Ham.)	*Rasbora daniconius*	*Cyprinus (Danio) daniconius*	*Dendua*
Crossocheilus latius latius (Ham.)	*Cirrhina latia*	*Cyprinus (Garr) latius*	*Patphorni*
Garra gotyla gotyla (Ham.)	*Discognathus lamta*	*Cyprinus lamta*	*Siltoka*
Labeo boga (Ham.)	*Labeo boga*	*Cyprinus (Bangana) boga*	*Bhagan*
Labeo dero (Ham.)	*Labeo microphthalmus*	*Cyprinus (Bangana) dero*	*Arangi*
Labeo gonius (Ham.)	*Labeo gonius*	*Cyprinus (Cyprinus) gomus*	*Kursi bata*
Labeo rohita (Ham.)	*Labeo rohita*	–	*Rohu*
Tor putitora (Ham.)	*Barbus putitora*	*Cyprinus (Cyprinus) putitora*	*Mahseer*
Tor tor (Ham.)	*Barbus tor*	*Cyprinus (Cyprinus) tor*	*Mahseer*
Puntius sarana sarana (Ham.)	*Barbus sarana*	*Cyprinus (Cyprinus) sarana*	*Darahee*
Puntius sophore (Ham.)	*Sarbus sophore*	*Cyprinus (Puntius) sophore*	*Sidhari*
Puntius ticto (Ham.)	*Sarbus ticto*	*Cyprinus (Puntius) ticto*	*Sidhari*

Contd...

Table 3–Contd...

Present Scientific Name	*Name According to Day*	*Name According to Hamilton*	*Local Name*
Schizothorax plagiostomus	*Orienus plagiostomus*	–	*Dhibru*
Schizothorax sinuatus	*Orienus sinuatus*	–	*Maseen*
Schizothoraichthys progastus	*Schizothorax progastus*	–	*Chonga*
Botia dario (Ham.)	*Botia dario*	*Cabitus geto*	*Baghauna*
Nemacheilus beavani (Gunther)	*Nemacheilus beavani*	–	*Natwa*
Nemacheilus botia (Ham.)	*Nemacheilus botia*	*Cabitus botia*	*Natwa*
Nemacheilus montanus (McClelland)	*Nemacheilus denisoni*	–	*Natwa*
Nemacheilus savona (Ham.)	*Nemacheilus savona*	*Cabitus turio*	*Natwa*
Mystus tengara (Ham.)	*Macrones lengara*	*Pimelodus tengara*	*Tengara*
Rita rita (Ham.)	*Rita buchanani*	*Pimelodus rita*	*Hunna rita*
Bagarius bagarius (Ham.)	*Bagarius yarrelli*	*Pimelodus bagarius*	*Gonch*
Glypthothorax pectinopterusrus (McClelland)	*Glypthothorax pectinopterusus*	–	*Nau*
Clupisoma garua (Ham.)	*Pseudeutropius garua*	*Silurus (Callichrous) garua*	*Salkari karahi*
Xenentoden cancila (Ham.)	*Belone cancila*	*Esor cancila*	*Kanwa*
Channa gachua (Ham.)	*Ophiocephalus gachua*	*Ophiocephalus gachua*	*Chanaga*
Colisa fasciatus (Block et Schn)	*Trichogaster fasciatus*	*Tricopodus colisa*	*Khosti*
Mastacembelus armatus (Lacepede)	*Mastacembelus armatus*	*Macrognathus armatus*	*Baam*

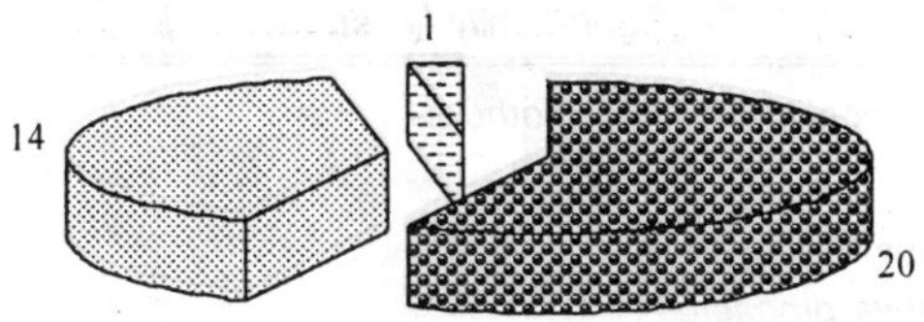

Figure 6(c): Distributional Pattern of Fish Species at Saptsarovar site in Ganga River

Table 4: Ichthyofaunic Diversity in Different Sampling Sites of Ganga River

Fish Species	*Family/ Sub-family*	*Up-stream*	*Down-stream*	*Sapt-sarovar*
Barilius bendelisis	**Cyprinidae**	a	c	a
Barilius bola	Rasborinae	a	c	c
Barilius vagra		a	r	r
Danio devario		a	c	c
Danio (Brachydanio) rorio		c	r	c
Esomus danricus		c	c	a
Rasbora daniconius		a	r	c
Crossocheilus latius latius	Garrinae	c	c	c
Garra gotyla gotyla		c	a	a
Labeo boga	*Cyprininae*	c	c	c
Labeo dero		a	r	c
Labeo gonius		c	r	a
Labeo rohita		c	a	a
Tor putitora		a	c	a
Tor tor		a	r	c
Puntius sarana sarana		c	a	a
Puntius sophore		c	a	a
Puntius ticto		r	c	a

Contd...

Table 4–Contd...

Fish Species	*Family/ Sub-family*	*Up-stream*	*Down-stream*	*Sapt-sarovar*
Schizothorax plagiostomus	*Schizothora-cinae*	a	c	c
Schizothorax sinuatus		a	r	c
Schizothoraichthys progastus		a	r	c
Botia dario	***Cobitidae/ Botinae***	c	c	a
Nemacheilus beavani	*Nemacheilinae*	c	a	a
Nemacheilus botia		c	a	a
Nemacheilus montanus		c	r	c
Nemacheilus savona		a	c	a
Mystus tengara	**Bagridae**	c	c	c
Rita rita		c	r	c
Bagarius bagarius	**Sisoridae**	c	r	a
Glyptothorax pectinopterusrus			r	a
Clupisoma garua	**Schilbeidae**	r	a	a
Xenentodon cancila	**Belonidae**	c	a	a
Ophiocephalus channa gachua	**Ophiocephali-dae**	r	r	a
Colisa fasciatus	**Anabantidae**	c	c	a
Mastacembelus armatus	**Mastacemeli-dae**	c	c	a

a: Abundance; c: Common; r: Rare.

Fish Catch Composition

Fish catch composition studies have revealed an important aspect of fishery biology. Selection of a population or stock of a particular fish species has shown a specific measurement to determine its distribution and size composition. Catch size of a fish species is an essential aspect for a fish stock in a particular water body. Therefore in the present study, catch composition of mahseers (*Tor tor* and *Tor putitora*) were recorded based on landing of local fisherman of Rishikesh, private anglers and fishermen's society of Ajeetpur section in Hardwar region (Table 5).

According to present data, *Schizothorax, Tor* and *Labeo* species are commonly found in upstream of barrage. The total fish catch were recorded highest during summer 68–148 kg followed by 89–118 kg in winter and 72–91 kg in rainy season of the year. The catch of mahseer was observed maximum 58.006 per cent in winter season and minimum 37.262 per cent in rainy season (Table 6 and Figure 7a).

Table 5: Seasonal Abundance and Distribution of Mahseer Species, *Tor tor* and *Tor putitora* in Ganga River

Sampling Stations	*Seasons*	*Tor tor*	*Tor putitora*
Upstream	Summer	++	++
	Rainy	++	++
	Winter	+++	+++
Downstream	Summer	++	++
	Rainy	++	++
	Winter	++	+++
Saptsarovar	Summer	++	++
	Rainy	+++	+++
	Winter	+++	+++

Table 6: Seasonal Catch Composition of Mahseer and Total Fish Landing from Upstream of Ganga River

Season	*Total Fish Catch Range (Kg.)*	*Fish Catch Mean Value (Kg.)*	*Mahseer Catch*	
			Weight (Kg.)	*Weight (%)*
Summer	68–148	118.000	58.654	49.706
Rainy	72–91	73.750	27.481	37.262
Winter	89–118	105.375	61.124	58.006

In the total mahseer catch, 351–500 mm size group were recorded dominant in total fish catch (Table 7 and Figure 7b). The maximum catch of largest size group 461–500 mm size of mahseer was recorded 47.61 per cent in winter months and minimum 8.592 per cent in rainy seasons of the year. In rainy season, the most exploited size

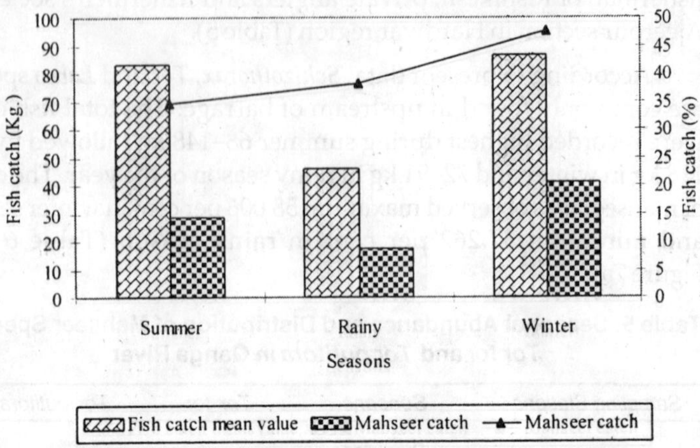

Figure 7(a): Seasonal Catch Composition of Mahseer and Total Fish Landing from Upstream of Ganga River

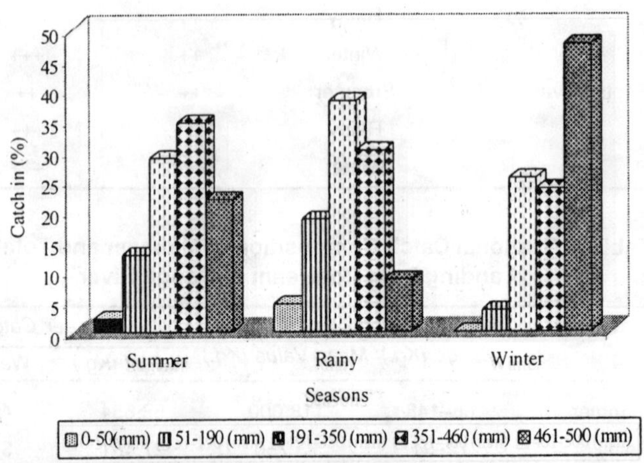

Figure 7(b): Seasonal Variations in the Percentage Frequencey of Different Size Group of Mahseer Catch from Upstream of Ganga River

group of mahseer was found 191 to 350 mm size, the percentage of this size group was observed 38.20 per cent (maximum) in this season. The mahseer of fingerling size 51–190 mm size groups were varied from 3.50 per cent in winter to 18.50 per cent in rainy season of throughout the year. The smallest size was upto 50 mm size, which caught due to illegal methods of fishing by the small section of fish anglers.

Table 7: Seasonal Variation in the Percentage Frequency of Different Size Groups of Mahseer Catch in Upstream of Ganga River

Season	*Length Group of Mahseer*				
	0–50 (mm)	*51–190 (mm)*	*191–350 (mm)*	*351–460 (mm)*	*461–500 (mm)*
Summer	2.20	12.75	28.654	34.522	21.874
Rainy	4.50	18.50	38.209	29.920	8.592
Winter	–	3.50	25.500	23.520	47.610

In the down stream of barrage, the total fish catch were recorded highest 43–168 kg in summer followed by 60–118 kg in winter and 46–76 kg in rainy season of the year. The highest mahseer catch was observed 41.070 per cent in winter months, whereas minimum catch were recorded 24.99 per cent in rainy season, when the total fish catch was observed highest 65.750 kgs (Table 8 and Figure 8a).

Table 8: Seasonal Catch Composition of Mahseer and Total Fish Landing from Downstream of Ganga River

Season	*Total Fish Catch Range (Kg.)*	*Fish Catch Mean Value (Kg.)*	*Mahseer Catch*	
			Weight (Kg.)	*Weight (%)*
Summer	43–168	65.750	23.236	35.339
Rainy	46–76	31.956	8.487	24.993
Winter	60–118	56.875	23.359	41.070

In the size range of mahseer catch, the most exploited size group was observed 191–350 mm and the maximum catch was recorded 49.50 per cent in summer months. The dominant size group was found 351–500 mm and the catch range was recorded as 5.32 per cent to 38.57 per cent in rainy to winter season of the year. The size group of 51–190 mm was observed maximum in 25.10 per cent in rainy and minimum 5.15 per cent were found in winter season. The smallest size group of mahseer fingerlings upto 50 mm was highest (6.20 per cent) in rainy season (Table 9 and Figure 8b).

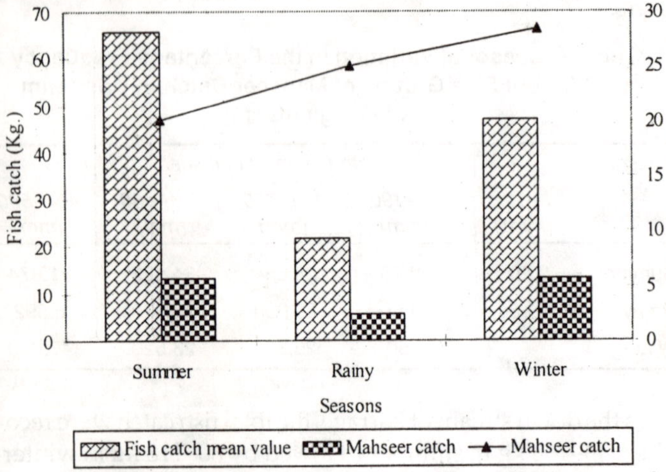

Figure 8(a): Seasonal Catch Composition of Mahseer and Total Fish Landing from Downstream of Ganga River

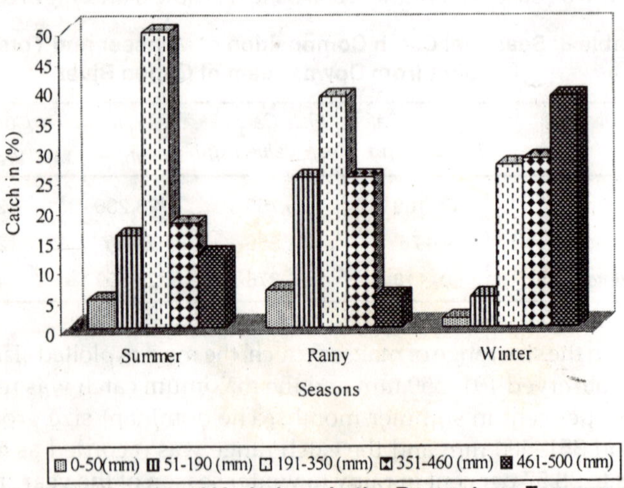

Figure 8(b): Seasonal Variations in the Percentage Frequencey of Different Size Group of Mahseer Catch from Downstream of Ganga River

Table 9: Seasonal Variation in the Percentage Frequency of Different Size Groups of Mahseer Catch in Downstream of Ganga River

Season	*Length Group of Mahseer*				
	0–50 (mm)	*51–190 (mm)*	*191–350 (mm)*	*351–460 (mm)*	*461–500 (mm)*
Summer	—	43–168	65.750	23.236	35.339
Rainy	—	46–76	31.956	8.487	24.993
Winter	—	60-118	56.875	23.359	41.070

Biometry

Tor tor

In the upstream of barrage, the total length range from 32.0 to 44.0 cm and weight range from 300.00 to 800.00 gms of fish specimens were used for morphometric study (Table 10 and Plate 6). The correlation coefficient 'r' was found to vary 0.72843 (LRB) to 0.99672 (SL) respectively with total length (TL) of fish. Standard length (SL, 0.99672) and head length (HL, 0.98844) were found the most highly correlated, while length of rostral barbel (LRB, 0.72843) and snout length (SnL, 0.77509) were observed to be the leasts correlated body parts (Table 11). The 'r' value of snout length (SnL, 0.65182) was least correlated and 'r' value of Inter orbital distance (IOD, 0.94906) was highly correlated with head length (HL) of fish. The regression coefficient 'b' ranged from 0.00042 (LRB) to 0.82761 (SL) respectively. The maximum 'b' value was recorded for SL (0.82761) followed by MW (0.43324). The 'b' value for LRB (0.00042) and ED (0.00050)

Plate 6: Deep bodied Mahseer, *Tor tor*

Table 10: Value of Morphometric Character of Mahseer, *Tor tor* in Upstream of Ganga River

Parameters	*Values±SD*	*Range*	*Range Difference*
Weight (gms.)	427.50±140.347	300.00–800.00	500.00
Total Length (cm.)	35.5±3.628	32.00–44.00	12.00
Standard Length (cm.)	29.27±3.301	26.10–36.80	10.70
Head Length (cm.)	7.22±0.458	6.80–8.20	1.40
Snout Length (cm.)	3.20±0.300	2.70–3.80	1.10
Max. Width (cm.)	17.25±1.845	14.10–20.20	6.10
Max. Girth (cm.)	5.34±0.534	4.80–6.10	1.30
Max. Height (cm.)	8.00±0.559	7.30–9.10	1.80
Eye Diameter (cm.)	1.30±0.092	1.20–1.50	0.30
Inter Orbital Distance (cm.)	3.21±0.233	3.00–3.80	0.80
Length of Rostral barbel (cm.)	1.175±0.084	1.00–1.30	0.30
Length of Maxillary barbel (cm.)	1.22±0.138	1.00–1.50	0.50
Length of Caudal fin (cm.)	7.50±0.557	7.00–8.90	1.90
Depth of Caudal fin (cm.)	3.42±0.609	3.00–5.10	2.10
Length of Dorsal fin (cm.)	5.83±0.330	5.50–6.60	1.10
Depth of Dorsal fin (cm.)	3.94±0.353	3.50–4.70	1.20
Length of Pectoral fin (cm.)	5.24±0.475	4.70–6.20	1.50
Length of Pelvic fin (cm.)	4.70±0.290	4.40–5.40	1.00
Length of Anal fin (cm.)	5.26±0.266	5.00–5.90	0.90

were observed to be least with total length (TL). The regression coefficient 'b' was maximum for SnL (0.28541) and minimum for LRB (0.02642) with head length (HL) of fish (Table 12 and Figures 10a–k).

In the downstream of barrage, the total range from 30.0 to 41.0 cm size and weight range form 250.0 to 600.00 gms of fishes were used for morphometric analysis (Table 14). The 'r' value was found to vary in the range from 0.70954 (SnL) to 0.99064 (SL) respectively with the total length of fish. Snout length (SnL, 0.70954) and eye diameter (ED, 0.80454) were the least correlated, while standard length (SL, 0.99064) and head length (HL, 0.95157) were the most highly correlated with total length (TL). With the head length the 'r'

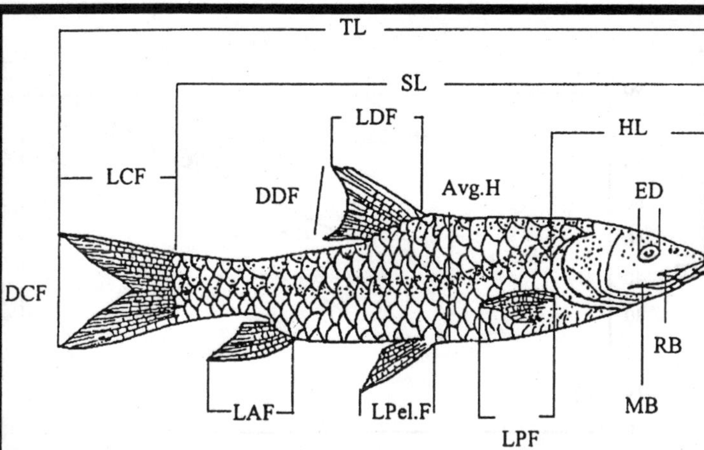

Figure 9: Showing Morphometric Characters Used for Biometric Studies of Mahseer Fishes

TL: Total length; SL: Standard length; HL: Head length; ED: Eye diameter; LCF: Lengh of caudal fin; DCF: Depth of caudal fin; LDF: Length of dorsal fin; DDF: Depth of dorsal fin; LAF: Length of anal fin; LPel.F: Length of pelvic fin; LPF: Length of pectoral fin; Avg. H: Average hegith; RB: Rostral barbell; MB: Maxillary barbel.

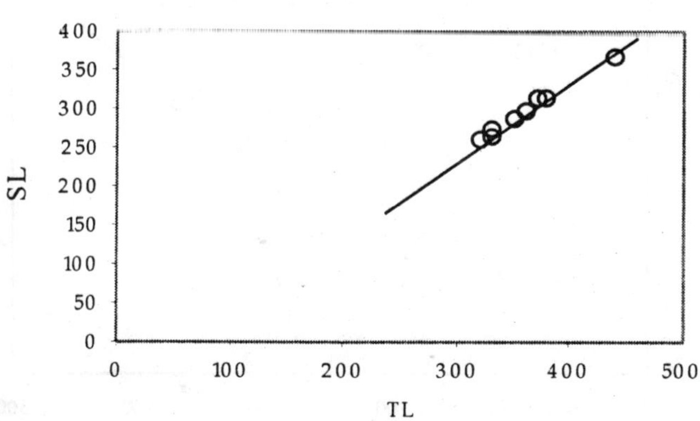

Figure 10(a): Linear Regression Between Total Length (TL) and Standard Length (SD) of *Tor tor* in Upstream of Ganga River

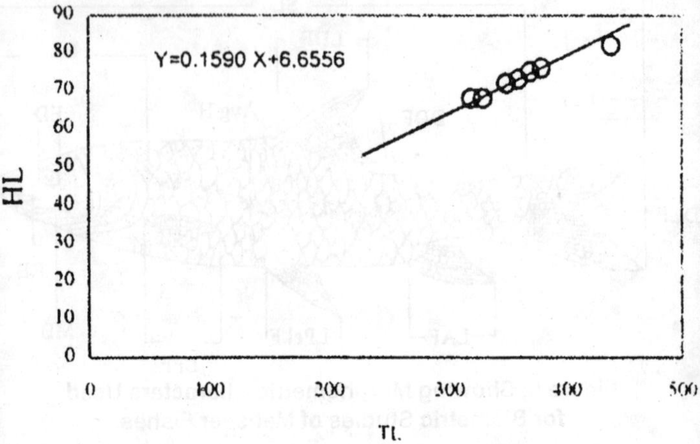

Figure 10(b): Linear Regression Between Total Length (TL) and Head Length (HD) of *Tor tor* in Upstream of Ganga River

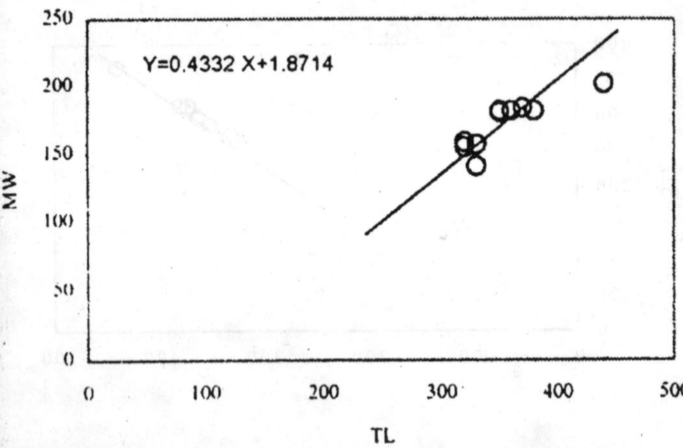

Figure 10(c): Linear Regression Between Total Length (TL) and Max. Width (MW) of *Tor tor* in Upstream of Ganga River

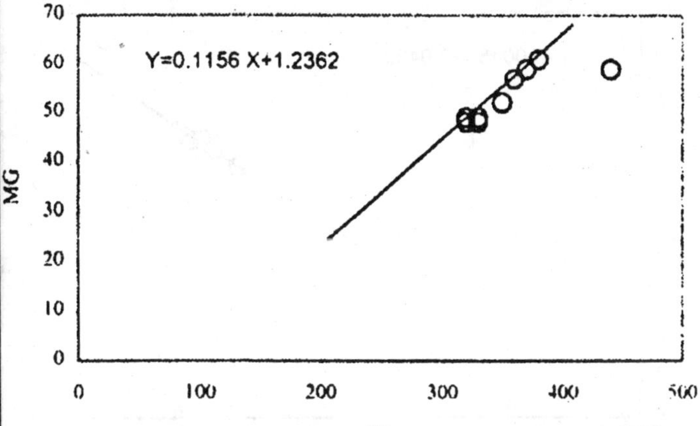

Figure 10(d): Linear Regression Between Total Length (TL) and Max. Girth (MG) of *Tor tor* in Upstream of Ganga River

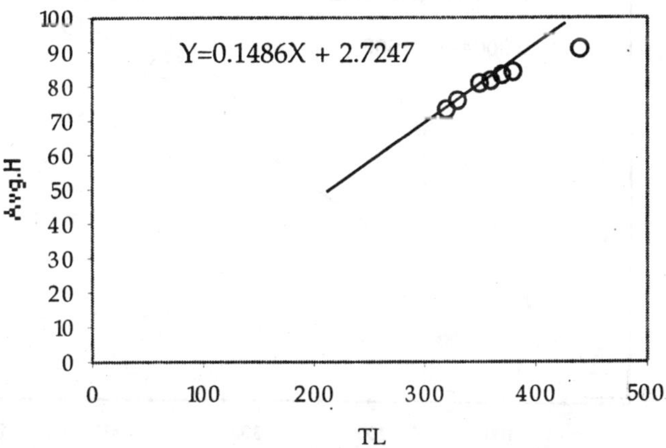

Figure 10(e): Linear Regression Between Total Length (TL) and Avg. H (Avg. H) of *Tor tor* in Upstream of Ganga River

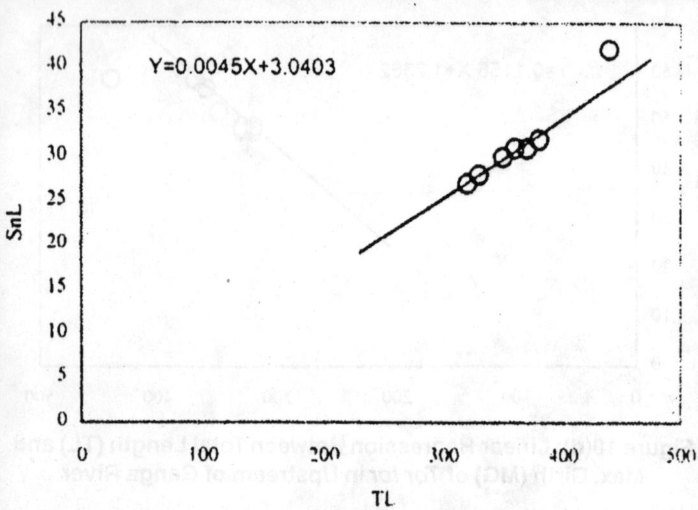

Figure 10(f): Linear Regression Between Total Length (TL) and Snout Length (SnL) of *Tor tor* in Upstream of Ganga River

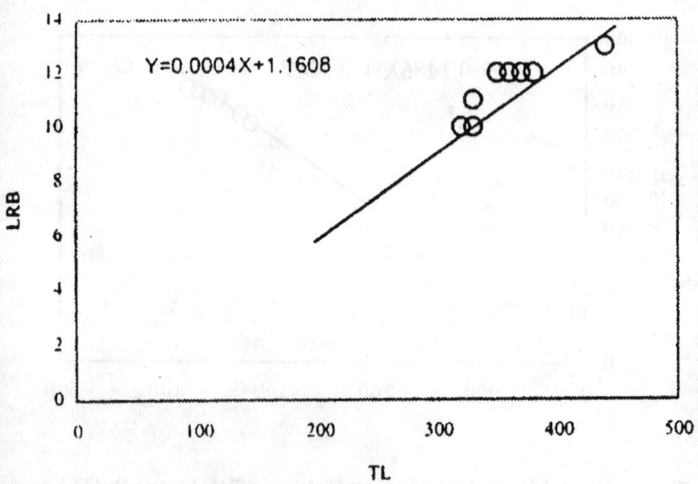

Figure 10(g): Linear Regression Between Total Length (TL) and Length of Rostral Barbel (LRB) of *Tor tor* in Upstream of Ganga River

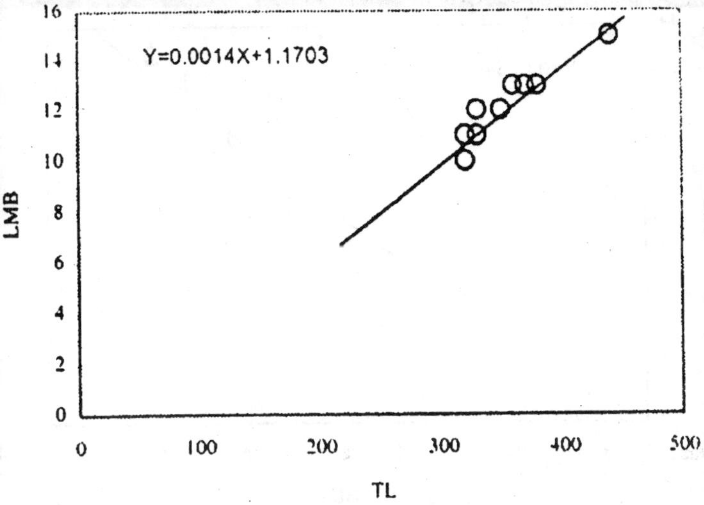

Figure 10(h): Linear Regression Between Total Length (TL) and Length of Maxillary Barbel (LMB) of *Tor tor* in Upstream of Ganga River

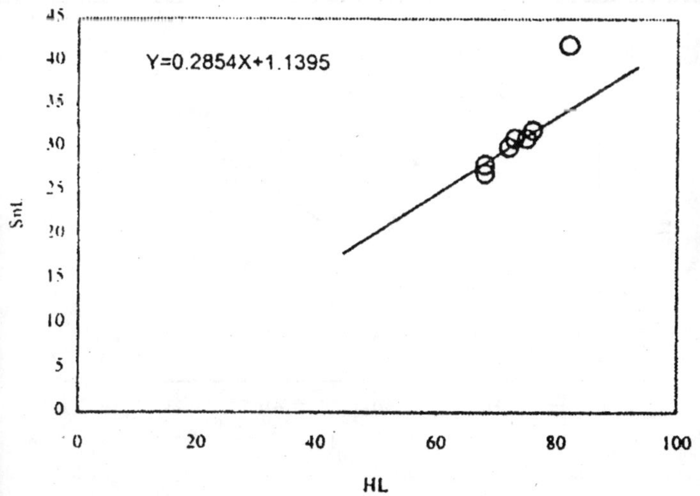

Figure 10(i): Linear Regression Between Head Length (HL) and Snout Length (SnL) of *Tor tor* in Upstream of Ganga River

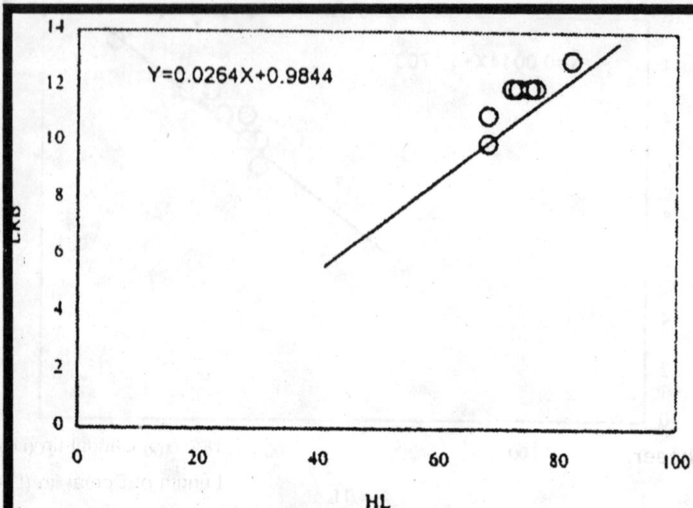

Figure 10(j): Linear Regression Between Head Length (HL) and Length of Rostral Barbel (LRB) of *Tor tor* in Upstream of Ganga River

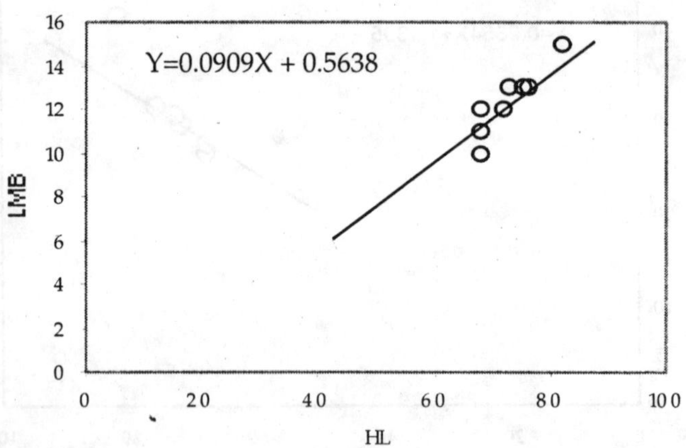

Figure 10(k): Linear Regression Between Head Length (HL) and Length of Maxillary Barbel (LMB) of *Tor tor* in Upstream of Ganga River

Table 11: Morphometric Analysis of Mahseer, *Tor tor* - Correlation of Various Body Parts with Total Length and Head Length in Upstream of Ganga River

In Relation to Total Length (TL)	*r–Value*
Standard Length (SL)	0.99672
Head Length (HL)	0.98844
Max. Width (MW)	0.85169
Max Girth (MG)	0.78504
Avg. Height (Avg. H)	0.96387
Length of Caudal fin (LCF)	0.98818
Depth of Caudal fin (DCF)	0.94498
Length of Dorsal fin (LDF)	0.93734
Depth of Dorsal fin (DDF)	0.86174
Length of Pectoral fin (LPF)	0.97812
Length of Pelvic fin (LPel.F)	0.91676
Length of Anal fin (LAP)	0.89566
Snout Length (SnL)	0.77509
Eye Diameter (ED)	0.82520
Inter Orbital Distance (IOD)	0.95931
Length of Rostral barbel (LRB)	0.72843
Length of Maxillary barbel (LMB)	0.95259
In Relation to Head length (HL)	*r–Value*
Snout Length (SnL)	0.65182
Eye Diameter (ED)	0.83086
Inter Orbital Distance (IOD)	0.94906
Length of Rostral barbel (LRB)	0.79478
Length of Maxillary barbel (LMB)	0.93684

value of LRB (0.97752) was highly correlated, while the 'r' of SnL (0.46654) was least correlated body parts (Table 15). The regression coefficient 'b' was varied from 0.00040 (LRB) to 0.94577 (SL) respectively with total length (TL) of fish and the least value of 'b' was recorded for (LRB). The maximum value was recorded for SL (0.94577) followed by Avg. H (0.54805). The 'b' values for LRB (0.00040) and LMB (0.00084) were observed to be least (Table 16 and Figures 11a–k). The 'b' value was recorded highest (0.50261) for IOD and least (0.01076) for LRB with head length (HL) of *Tor tor*.

Table 12: Recial Analysis of Mahseer, *Tor tor* in Upstream of Ganga river–Regression Coefficient and Equation of Various Body Parts in Relation to Total Length and Head Length

Body Parameters	*b–Value with Total Length (TL)*	
	Regression Coefficient	*Regression Equation*
Standard Length (SL)	0.82761	Y = 0.8276 X + 0.1098
Head Length (HL)	0.01596	Y = 0.1590 X + 6.6556
W Avg. Height (Avg. H)	0.14869	Y = 0.1486 X + 2.7247
Max. Width (MW)	0.43324	Y = 0.4332 X + 1.8714
Max. Girth (MG)	0.11561	Y = 0.1156 X + 1.2362
Length of Caudal fin (LCF)	0.15189	Y = 0.1518 X + 2.1111
Depth of Caudal fin (DCF)	0.15864	Y = 0.1586 X – 2.2103
Length of Dorsal fin (LDF)	0.00739	Y = 0.0073 X + 5.5709
Depth of Dorsal fin (DDF)	0.00737	Y = 0.0073 X + 3.6809
Length of Pectoral fin (LPF)	0.01719	Y = 0.0171 X + 4.6330
Length of Pelvic fin (LPel.F)	0.07341	Y = 0.0734 X + 2.0943
Length of Anal fin (LAF)	0.06582	Y = 0.0658 X + 2.9241
Snout Length (SnL)	0.00455	Y = 0.0045 X + 3.0403
Eye Diameter (ED)	0.00050	Y = 0.0005 X + 1.2823
Inter Orbital Distance (IOD)	0.06168	Y = 0.0616 X + 1.0232
Length of Rostral barbel (LRB)	0.00042	Y = 0.0004 X + 1.1608
Length of Maxillary barbel (LMB)	0.00145	Y = 0.0014 X + 1.1703
Body Parameters	*b–Value with Head Length (HL)*	
	Regression Coefficient	*Regression Equation*
Snout Length (SnL)	0.28541	Y = 0.2854 X + 1.1395
Eye Diameter (ED)	0.03171	Y = 0.0317 X + 1.0721
Inter Orbital Distance (IOD)	0.25898	Y = 0.2589 X + 1.3408
Length of Rostral barbel (LRB)	0.02642	Y = 0.0264 X + 0.9844
Length of Maxillary barbel (LMB)	0.09090	Y = 0.0909 X + 0.5638

Tor putitora

In the upstream, total length range from 28.0 to 57.30 cm and weight range from 210 to 1800 gms of specimens were selected for

Table 13: Meristic Characters of Mahseer, *Tor tor* in Upstream of Ganga River

Characters	*Range*	*Mode*
Lateral line scale	20–27	25
Lateral line transverse scale	4½–4½	4½–4½
Pectoral fin rays	17–19	17
Dorsal fin rays	10–12 (2–3/9)	11
Caudal fin rays	19–20	19
Anal fin rays	7–8	7
Ventral fin rays	9–10	9

Table 14: Value of Different Morphometric Character of Mahseer, *Tor tor* in Downstream of Ganga River

Parameters	*Values±SD*	*Range*	*Range Difference*
Weight (gms.)	335.0±121.220	250.00–600.00	350.00
Total Length (cm.)	33.90±3.204	30.00–41.00	11.00.
Standard Length (cm.)	27.02±3.059	23.10–33.40	10.30
Head Length (cm.)	6.56±0.618	5.50–7.60	2.10
Snout Length (cm.)	3.08±0.228	2.50–3.40	0.90
Max. Width (cm.)	16.59±1.901	13.90–19.40	5.50
Max. Girth (cm.)	5.10±0.365	4.60–5.70	1.10
Max. Height (cm.)	7.14±0.682	6.10–7.80	1.70
Eye Diameter (cm.)	1.26±0.128	1.10–1.50	0.40
Inter Orbital Distance (cm.)	3.31±0.438	2.30–3.80	1.50
Length of Rostral barbel (cm.)	1.14±0.068	1.00–1.20	0.20
Length of Maxillary barbel (cm.)	1.34±0.093	1.20–1.50	0.30
Length of Caudal fin (cm.)	7.11±0.375	6.50–7.70	1.20
Depth of Caudal fin (cm.)	3.30±0.274	2.90–3.80	0.90
Length of Dorsal fin (cm.)	5.74±0.230	5.40–6.20	0.80
Depth of Dorsal fin (cm.)	4.08±2.273	3.70–4.50	0.80
Length of Pectoral fin (cm.)	5.32±0.252	5.00–5.80	0.80
Length of Pelvic fin (cm.)	4.48±0.345	4.10–5.20	1.10
Length of Anal fin (cm.)	5.18±0.436	4.60–5.80	1.20

Table 15: Morphometric Analysis of Mahseer, *Tor tor* - Correlation of Various Body Parts with Total Length and Head Length in Downstream of Ganga River

In Relation to Total Length (TL)	*r–Value*
Standard Length (SL)	0.99064
Head Length (HL)	0.95157
Avg. Height (Avg. H)	0.92369
Max. Width (MW)	0.93543
Max. Girth (MG)	0.95151
Length of Caudal fin (LCF)	0.88321
Depth of Caudal fin (DCF)	0.85156
Length of Dorsal fin (LDF)	0.92389
Depth of Dorsalfin (DDF)	0.84139
Length of Pectoral fin (LPF)	0.86246
Length of Pelvic fin (LPel.F)	0.93745
Length of Anal fin (LAP)	0.83324
Snout Length (SnL)	0.70954
Eye Diameter (ED)	0.80454
Inter Orbital Distance (IOD)	0.82384
Length of Rostral barbel (LRB)	0.82467
Length of Maxillary barbel (LMB)	0.94634
In Relation to Head length (HL)	*r–Value*
Snout Length (SnL)	0.46654
Eye Diameter (ED)	0.74363
Inter Orbital Distance (IOD)	0.93475
Length of Rostral barbel (LRB)	0.97752
Length of Maxillary barbel (LMB)	0.85361

morphometric analysis (Table 18 and Plate 7). The 'r' value was found to vary from 0.53483 (SnL) to 0.99803 (SL) respectively with total length (TL). The 'r' value was calculated as 0.39058 (SnL) to 0.95798 (IOD) with head length (HL) of fish (Table 19). Standard length (SL, 0.99803) and head length (HL, 0.98897) were found the highly correlated. While, snout length (SnL, 0.53483) and length of maxillary barbel (LMB, 0.86572) were observed to be the least correlated body parameters with total length of fish. The regression

Table 16: Recial Analysis of Mahseer, *Tor tor* in Downstream of Ganga River–Regression Coefficient and Equation of Various Body Parts in Relation Total Length and Head Length

Body Parameters	*b–Value with Total Length (TL)*	
	Regression Coefficient	*Regression Equation*
Standard Length (SL)	0.94577	Y = 0.9457 X – 5.0392
Head Length (HL)	0.03720	Y = 0.0372 X + 5.2990
Avg. Height (Avg. H)	0.04543	Y = 0.0454 X + 5.6010
Max. Width (MW)	0.54805	Y = 0.5480 X – 1.9872
Max. Girth (MG)	0.10660	Y = 0.1066 X + 1.4863
Length of Caudal fin (LCF)	0.10346	Y = 0.1034 X + 3.6048
Depth of Caudal fin (DCF)	0.07305	Y = 0.0730 X + 0.8253
Length of Dorsal fin (LDF)	0.06645	Y = 0.0664 X + 3.4891
Depth of Dorsal fin (DDF)	0.07175	Y = 0.0717 X + 1.6494
Length of Pectoral fin (LPF)	0.06785	Y = 0.0678 X + 3.0216
Length of Pelvic fin (LPel.F)	0.10097	Y = 0.1009 X + 1.0595
Length of Anal fin (LAF)	0.11341	Y = 0.1134X + 1.3358
Snout Length (SnL)	0.00396	Y = 0.0039 X + 2.9478
Eye Diameter(ED)	0.00124	Y = 0.0012 X + 1.2214
Inter Orbital Distance (IOD)	0.11266	Y = 0.1126 X – 0.5071
Length of Rostral barbel (LRB)	0.00040	Y = 0.0004 X + 1.1265
Length of Maxillary barbel (LMB)	0.00084	Y = 0.0008 X + 1.3129
Body Parameters	*b–Value with Head Length (HL)*	
	Regression Coefficient	*Regression Equation*
Snout Length (SnL)	0.1.0645	Y = 0.1064 X + 2.3821
Eye Diameter (ED)	0.03344	Y = 0.0334 X + 1.0429
Inter Orbital Distance (IOD)	0.50261	Y = 0.5026 X + 0.0130
Length of Rostral barbel (LRB)	0.01076	Y = 0.0107 X + 1.0699
Length of Maxillary barbel (LMB)	0.02268	Y = 0.0226 X + 1.1918

cofficient 'b' ranged from 0.00692 (Avg. H) to 0.88281 (SL) respectively. The maximum value was recorded for SL (0.88281) followed by MW (0.32809). The 'b' values for Avg. H (0.00692) and SnL (0.01139) were observed to be least (Table 20 & Figures 12a-k). The b value was calculated highest (0.55714) for IOD and least (0.03895) for SnL with Head length of fish.

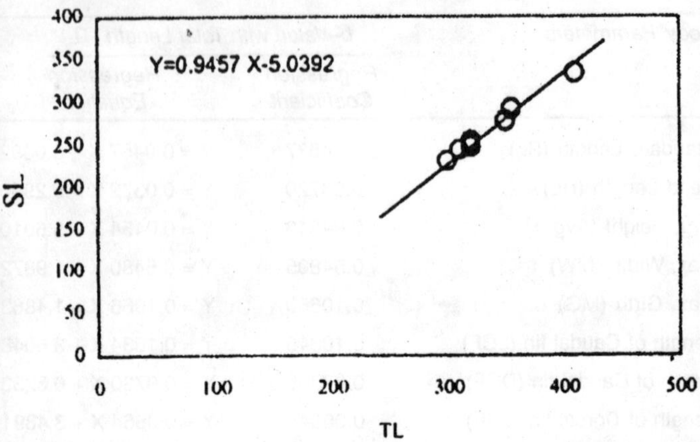

Figure 11(a): Linear Regression Between Total Length (TL) and Standard Length (SL) of *Tor tor* in Downstream of Ganga River

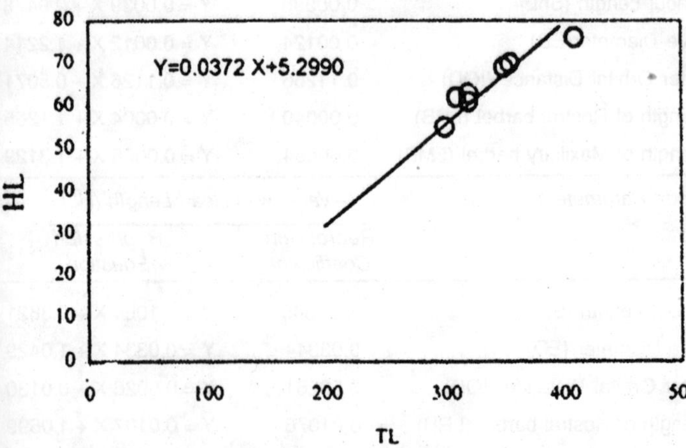

Figure 11(b): Linear Regression Between Total Length (TL) and Head Length (HL) of *Tor tor* in Downstream of Ganga River

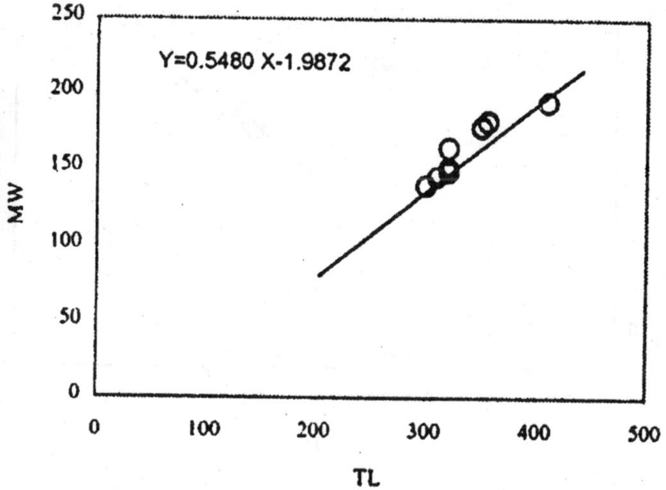

Figure 11(c): Linear Regression Between Total Length (TL) and Max. Width (MW) of *Tor tor* in Downstream of Ganga River

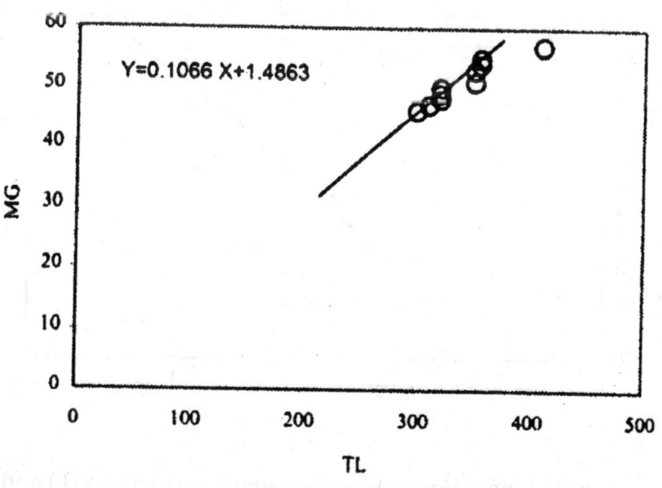

Figure 11(d): Linear Regression Between Total Length (TL) and Max. Girth (MG) of *Tor tor* in Downstream of Ganga River

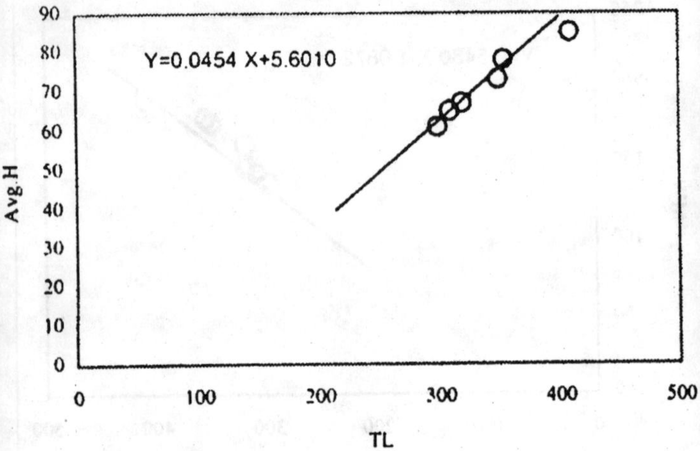

Figure 11(e): Linear Regression Between Total Length (TL) and Avg. H (Avg. H) of *Tor tor* in Downstream of Ganga River

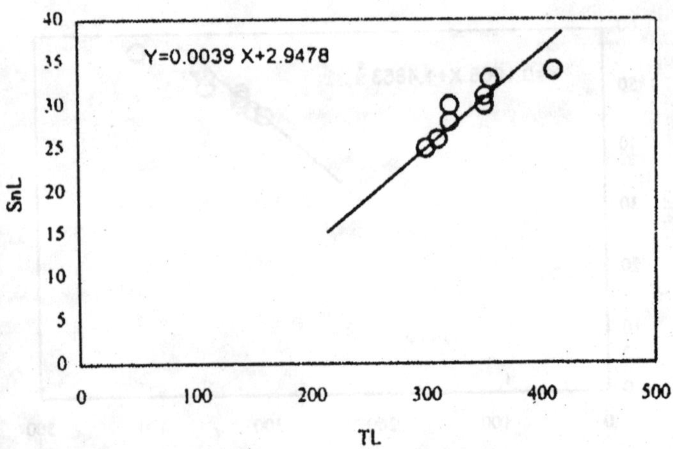

Figure 11(f): Linear Regression Between Total Length (TL) and Snout Length (SnL) of *Tor tor* in Downstream of Ganga River

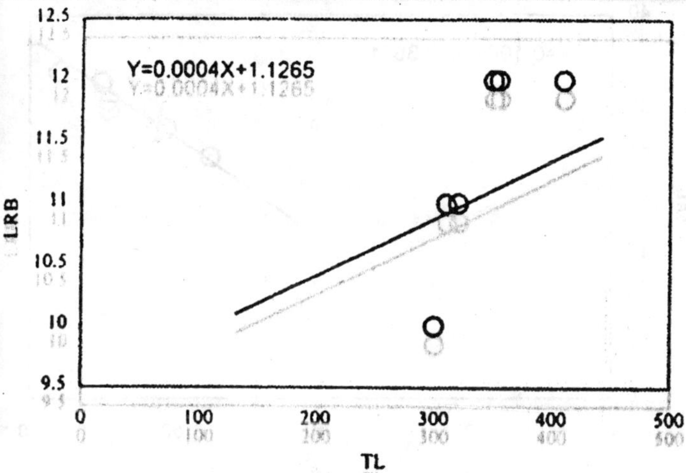

Figure 11(g): Linear Regression Between Total Length (TL) and Length of Rostral Barbel (LRB) of *Tor tor* in Downstream of Ganga River

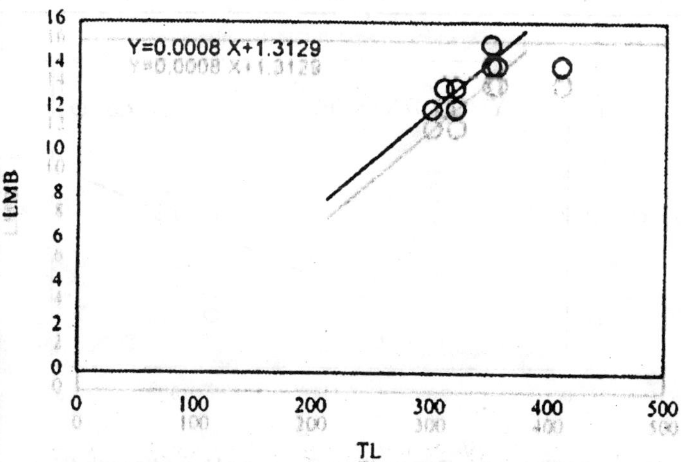

Figure 11(h): Linear Regression Between Total Length (TL) and Length of Maxillary Barbel (LMB) of *Tor tor* in Downstream of Ganga River

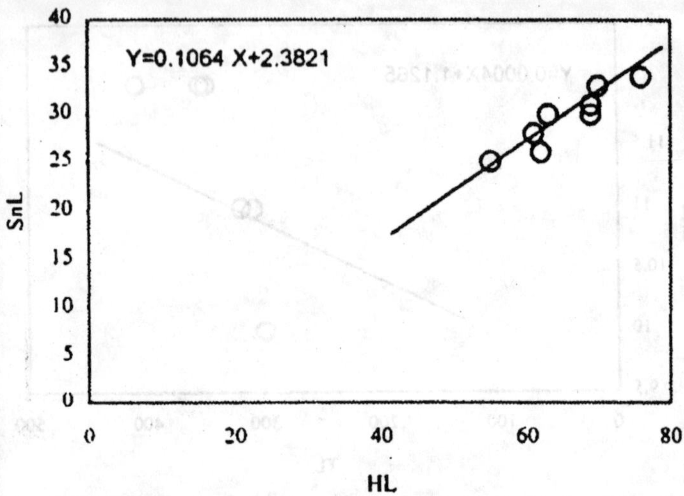

Figure 11(i): Linear Regression Between Head Length (HL) and Snout Length (SnL) of *Tor tor* in Downstream of Ganga River

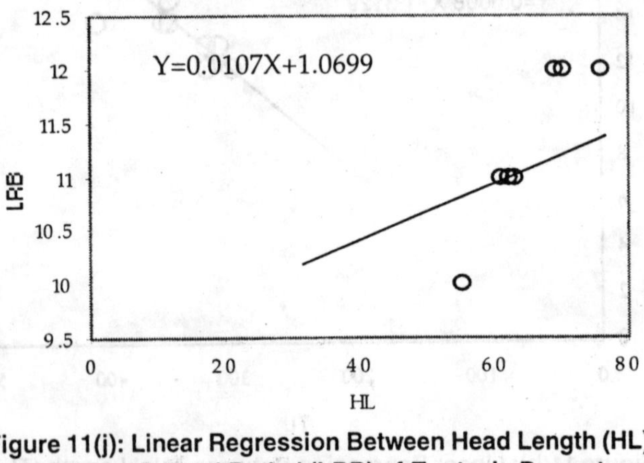

Figure 11(j): Linear Regression Between Head Length (HL) and Length of Rostral Barbel (LRB) of *Tor tor* in Downstream of Ganga River

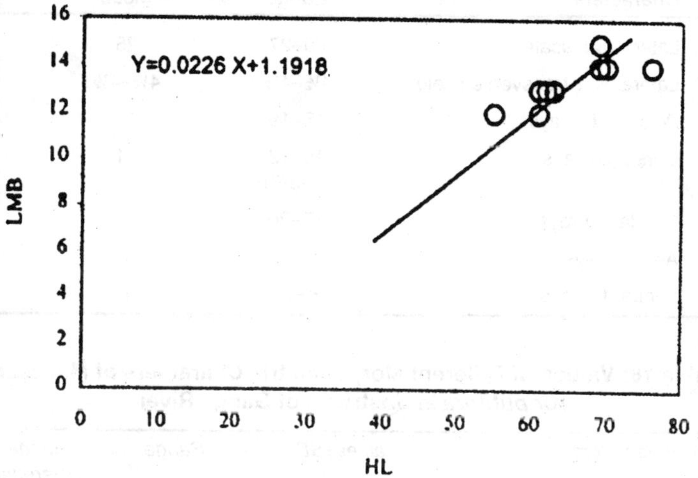

Figure 11(k): Linear Regression Between Head Length (HL) and Length of Maxillary Barbel (LMB) of *Tor tor* in Downstream of Ganga River

Plate 7: Himalayan Mahseer, *Tor putitora*

Table 17: Meristic Characters of Mahseer, *Tor tor* for Downstream of Ganga River

Characters	*Range*	*Mode*
Lateral line scale	20–27	25
Lateral line transverse scale	4½–4½	4½–4½
Pectoral fin rays	15–19	17
Dorsal fin rays	10–12 (2–3/9)	11
Caudal fin rays	17–20	17
Anal fin rays	5–8 7	7
Ventral fin rays	8–10	9

Table 18: Values of Different Morphometric Characters of Mahseer, *Tor putitora* in Upstream of Ganga River

Parameters	*Values±SD*	*Range*	*Range Difference*
Weight (gms.)	667.50±449.992	210–1800	1590.0
Total Length (cm.)	40.125±8.002	28.0–57.30	29.30
Standard Length (cm.)	33.80±7.078	22.20–48.50	26.30
Head Length (cm.)	9.55±1.709	6.40–13.10	6.70
Snout Length (cm.)	3.80±0.170	3.60–4.10	0.50
Max. Width (cm.)	17.275±2.808	13.80–23.10	9.30
Max. Girth (cm.)	7.816±1.557	5.90–11.60	5.70
Avg. Height (cm.)	7.966±0.666	6.90–9.40	2.50
Eye Diameter (cm.)	1.35±0.296	1.10–2.10	1.00
Inter Orbital Distance (cm.)	4.10±0.994	2.40–5.80	3.40
Length of Rostral barbel (cm.)	2.10±0.351	1.70–2.90	1.20
Length of Maxillary Barbel (cm.)	2.175±0.312	1.90–2.90	1.00
Length of Caudal fin (cm.)	8.766±1.414	6.90–11.20	4.30
Depth of Caudal fin (cm.)	4.325±0.967	3.10–5.60	2.50
Length of Dorsal fin (cm.)	6.575±1.096	5.20–8.30	3.10
Depth of Dorsal fin (cm.)	4.191±1.054	3.30–6.50	3.20
Length of Pectoral fin (cm.)	6.041±1.050	4.50–8.00	3.50
Length of Pelvic fin (cm.)	5.333±1.110	3.90–7.60	3.70
Length of Anal fin (cm.)	5.508±0.969	4.10–7.30	3.20

Table 19: Morphometric Analysis of Mahseer, *Tor putitora*, Correlation of Various Body Parts with Total Length and Head Length in Upstream of Ganga River

In Relation to Total Length (TL)	*r–Value*
Standard Length (SL)	0.99803
Head Length (HL)	0.98897
Max. Width (MW)	0.93506
Max. Girth (MG)	0.90725
Avg. Height (Avg. H)	0.96348
Length of Caudal fin (LCF)	0.94285
Depth of Caudal fin (DCF)	0.89236
Length of Dorsal fin (LDF)	0.92189
Depth of Dorsal fin (DDF)	0.95765
Length of Pectoral fin (LPF)	0.97153
Length of Pelvic fin (LPel.F)	0.95019
Length of Anal fin (LAP)	0.94503
Snout Length (SnL)	0.53483
Eye Diameter (ED)	0.87133
Inter Orbital Distance (IOD)	0.96410
Length of Rostral barbel (LRB)	0.91587
Length of Maxillary barbel (LMB)	0.86572
In Relation to Head length (HL)	*r–Value*
Snout Length (SnL)	0.39058
Eye Diameter (ED)	0.85470
Inter Orbital Distance (IOD)	0.95798
Length of Rostral barbel (LRB)	0.91184
Length of Maxillary barbel (LMB)	0.85238

In downstream, the length range of 14.0 to 43.50 cm. and weight range 50.00 to 720.00 gms were recorded for morphometric analysis (Table 22). The 'r' value was found to be vary from 0.61108 (SnL) to 0.99723 (SL) with total length (TL) and 0.67069 (SnL) to 0.98825 (IOD) with head length (HL) of fish (Table 23). Standard length (SL, 0.99723) and head length (HL, 0.99298) were the most highly correlated body parts. Snout length (SnL, 0.61108) and length of

Table 20: Recial Analysis of Mahseer *Tor putitora* in Upstream of Ganga River–Regression Coefficient and Equation of Various Body Parts in Relation to Total Length and Head Length

Body Parameters	*b–Value with Total Length (TL)*	
	Regression Coefficient	*Regression Equation*
Standard Length (SL)	0.88281	Y = 0.8828 X – 1.6223
Head Length(HL)	0.21131	Y = 0.2113 X + 1.0716
Max. Width (MW)	0.32809	Y = 0.3280 X + 4.1140
Max. Girth (MG)	0.17658	Y = 0.1765 X + 0.7340
Avg. Height (Avg. H)	0.00692	Y = 0.0069 X + 7.6892
Length of Caudal fin (LCF)	0.16662	Y = 0.1666 X + 2.0812
Depth of Caudal fin (DCF)	0.10787	Y = 0.1078 X – 0.0032
Length of Dorsal fin (LDF)	0.12635	Y = 0.1263 X + 1.5073
Depth of Dorsal fin (DDF)	0.12614	Y = 0.1261 X – 0.8687
Length of Pectoral fin (LPF)	0.12755	Y = 0.1275 X + 0.9251
Length of Pelvic fin (LPel.F)	0.13183	Y = 0.1318 X + 0.0446
Length of Anal fin (LAF)	0.11450	Y = 0.1145 X + 0.9137
Snout Length (SnL)	0.01139	Y = 0.0113 X + 3.3466
Eye Diameter (ED)	0.03223	Y = 0.0322 X + 0.0580
Inter Orbital Distance (IOD)	0.11981	Y = 0.1198 X – 0.7069
Length of Rostral barbel (LRB)	0.04024	Y = 0.0402 X + 0.4870
Length of Maxillary barbel (LMB)	0.03381	Y = 0.0338 X + 0.8188
Body Parameters	*b–Value with Head Length (HL)*	
	Regression Coefficient	*Regression Equation*
Snout Length (SnL)	0.03895	Y = 0.0389 X + 3.4286
Eye Diameter (ED)	0.14796	Y = 0.1479 X + 0.0624
Inter Orbital Distance (IOD)	0.55714	Y = 0.5571 X – 1.2200
Length of Rostral barbel (LRB)	0.18747	Y = 0.1874 X + 0.3370
Length of Maxillary barbel (LMB)	0.15582	Y = 0.1558 X + 0.6880

rostral barbel (LRB, 0.84154) were observed to be the least correlated body parameters with total length (TL) of fish (Table 24 and Figures 13a–j). The regression coefficient 'b' was recorded

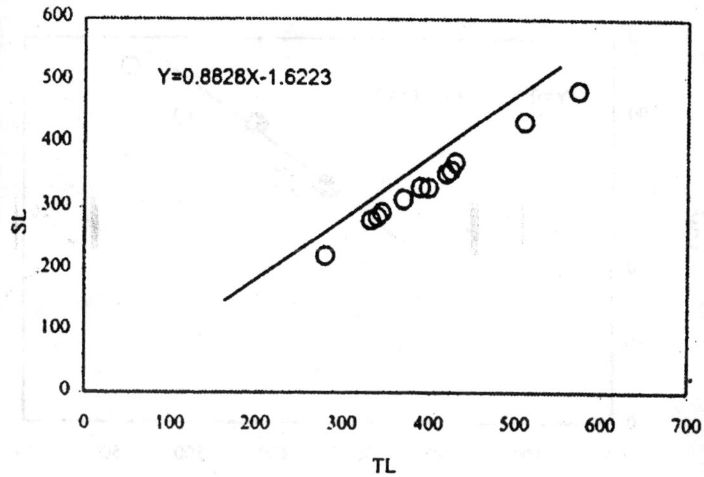

Figure 12(a): Linear Regression Between Total Length (TL) and Standard Length (SD) of *Tor putitora* in Upstream of Ganga River

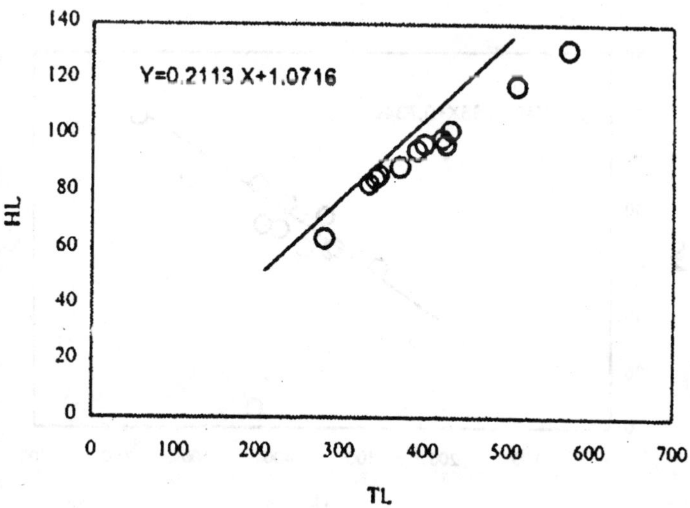

Figure 12(b): Linear Regression Between Total Length (TL) and Head Length (HD) of *Tor putitora* in Upstream of Ganga River

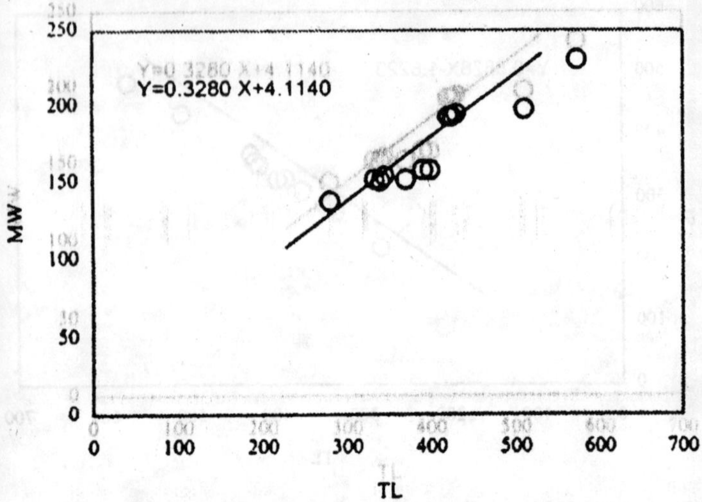

Figure 12(c): Linear Regression Between Total Length (TL) and Max. Width (MW) of *Tor putitora* in Upstream of Ganga River

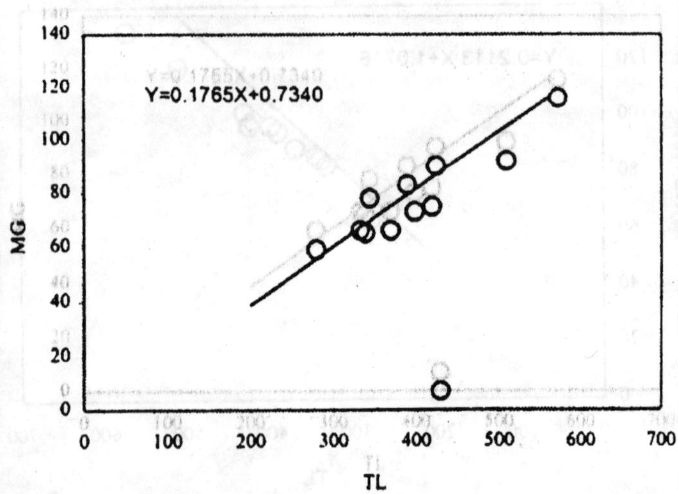

Figure 12(d): Linear Regression Between Total Length (TL) and Max. Girth (MG) of *Tor putitora* in Upstream of Ganga River

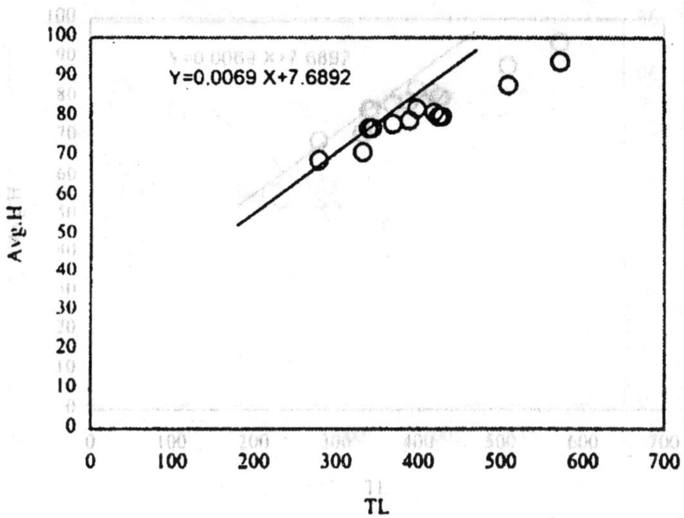

Figure 12(e): Linear Regression Between Total Length (TL) and Avg. H (Avg. H) of *Tor putitora* in Upstream of Ganga River

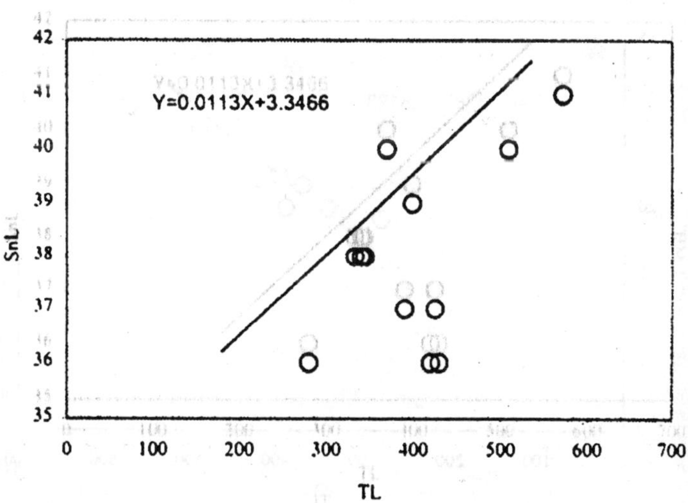

Figure 12(f): Linear Regression Between Total Length (TL) and Snout Length (SnL) of *Tor putitora* in Upstream of Ganga River

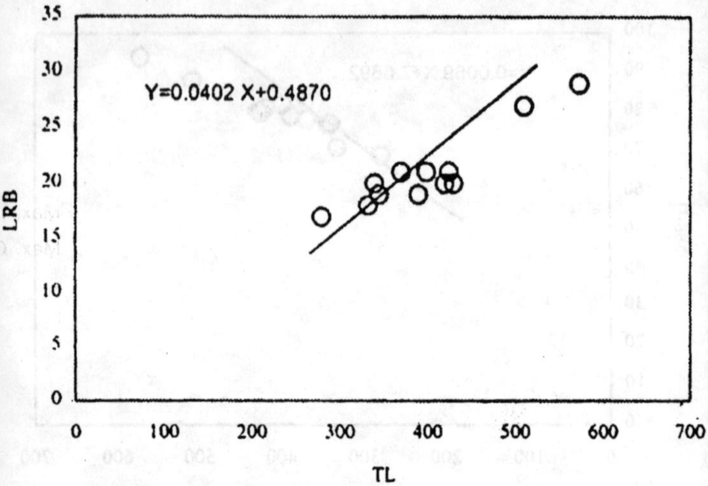

Figure 12(g): Linear Regression Between Total Length (TL) and Length of Rostral Barbel (LRB) of *Tor putitora* in Upstream of Ganga River

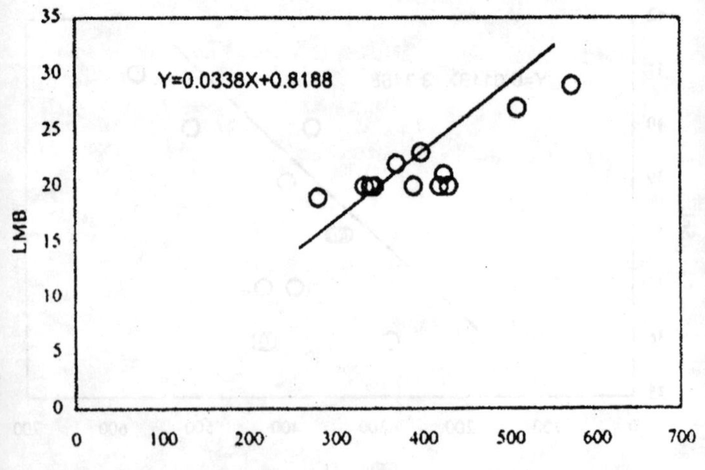

Figure 12(h): Linear Regression Between Total Length (TL) and Length of Maxillary Barbel (LMB) of *Tor putitora* in Upstream of Ganga River

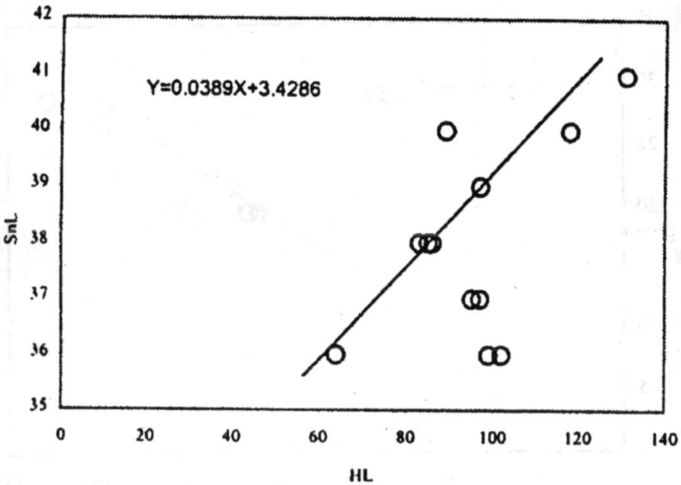

Figure 12(i): Linear Regression Between Head Length (HL) and Snout Length (SnL) of *Tor putitora* in Upstream of Ganga River

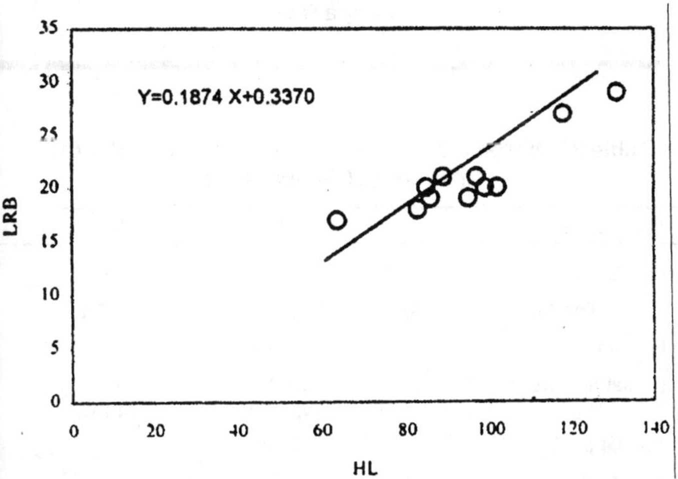

Figure 12(j): Linear Regression Between Head Length (HL) and Length of Rostral Barbel (LRB) of *Tor putitora* in Upstream of Ganga River

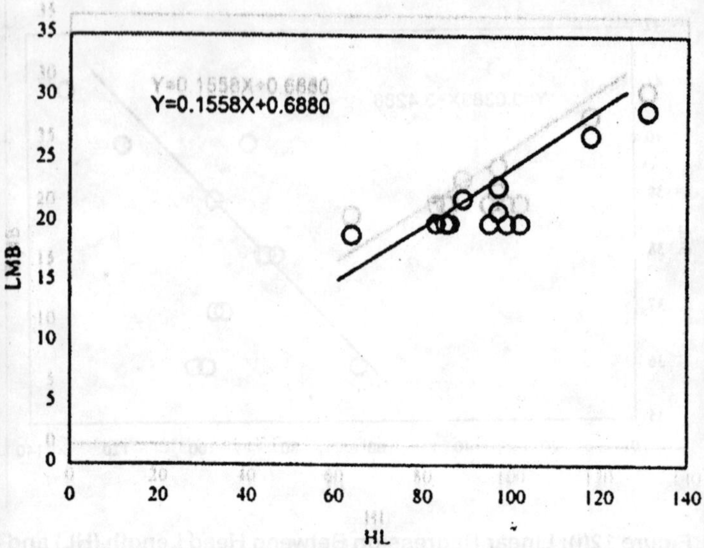

Figure 12(k): Linear Regression Between Head Length (HL) and Length of Maxillary Barbel (LMB) of *Tor putitora* in Upstream of Ganga River

Table 21: Meristic Characters of Mahseer, *Tor putitora* in Upstream of Ganga River

Characters	*Range*	*Mode*
Lateral line scale	24–28	27
Lateral line transverse scale	4½–3½	4½–3½
Pectoral fin rays	14–18	17
Dorsal fin rays	10–13 (2/9–4/9)	13 (4/9)
Caudal fin rays	19–22	20
Anal fin rays	5–8 (2/5–3/5)	7 (2/5)
Ventral fin rays	8–10	9

maximum for SL (0.81164) followed by HL (0.19735), the 'b' value was observed least (0.001) for ED and (0.00284) for LMB. The 'b' value was calculated least (0.06950) for LRB and highest (0.59363) for IOD was recorded with head length of fish.

Table 22: Values of Different Morphometric Characters of Mahseer, *Tor putitora* in Downstream of Ganga River

Parameters	*Values±SD*	*Range*	*Range Difference*
Weight (gms.)	400.50±220.837	50.00–720.00	670.00
Total Length (cm.)	34.00±9.064	14.00–43.50	29.50
Standard Length (cm.)	27.30±7.377	11.20–34.40	23.20
Head Length (cm.)	7.72±1.801	4.00–9.80	5.80
Snout Length (cm.)	4.21±0.648	2.70–5.10	2.40
Max. Width (cm.)	12.48±1.163	9.80–13.90	4.10
Max. Girth (cm.)	6.13±1.308	3.10–7.30	4.20
Max. Height (cm.)	6.14±1.237	3.40–7.40	4.00
Eye Diameter (cm.)	1.20±0.288	0.50–1.40	0.90
Inter Orbital Distance (cm.)	3.23±1.082	0.90–4.50	3.60
Length of Rostral barbel (cm.)	1.825±0.538	0.50–2.10	1.60
Length of Maxillary barbel (cm.)	1.855±0.512	0.50–2.10	1.60
Length of Caudal fin (cm.)	6.94±1.616	3.50-8.70	5.20
Depth of Caudal fin (cm.)	3.77±1.100	1.70-5.30	3.60
Length of Dorsal fin (cm.)	5.32±1.258	2.90–7.20	4.30
Depth of Dorsal fin (cm.)	3.48±0.814	1.80-4.50	2.70
Length of Pectoral fin (cm.)	4.92±1.116	3.00-6.50	3.50
Length of Pelvic fin (cm.)	4.48±1.181	2.00–5.90	3.90
Length of Anal fin (cm.)	4.37±1.316	1.90–6.00	4.10

The head length is greater than the depth of the body and is contained from 4.0 to 4.5 and depth of body 5.5 to 7 in the total length. The eye diameter was contained from 3.0 to 5.3 in the length of head and lies in the anterior half of the head. The depth of the body was contained from 1.1 to 1.4 times in the length of the head, 0.8 to 1.7 times in the length of the snout and 0.7 to 1.4 times in the

Table 23: Morphometric Analysis of Mahseer, *Tor putitora* - Correlation of Various Body Parts with Total Length and Head Length in Downstream of Ganga River

In Relation to Total Length (TL)	*r–Value*
Standard Length (SL)	0.99723
Head Length (HL)	0.99298
Max. Width (MW)	0.87236
Max. Girth (MG)	0.99230
Avg. Height (Avg. H)	0.89743
Length of Caudal fin (LCF)	0.88246
Depth of Caudal fin (DCF)	0.84971
Length of Dorsal fin (LDF)	0.93493
Depth of Dorsal fin (DD.F)	0.95643
Length of Pectoral fin (LPF)	0.91779
Length of Pelvic fin (LPel.F)	0.88340
Length of Anal fin (LAF)	0.94196
Snout Length (SnL)	0.61108
Eye Diameter (ED)	0.88453
Inter Orbital Distance (IOD)	0.99345
Length of Rostral barbel (LRB)	0.84154
Length of Maxillary barbel (LMB)	0.86713
In Relation to Head length (HL)	*r–Value*
Snout Length (SnL)	0.67069
Eye Diameter (ED)	0.88974
Inter Orbital Distance (IOD)	0.98825
Length of Rostral barbel (LRB)	0.79467
Length of Maxillary barbel (LMB)	0.83599

inter orbital distance. The least height of the caudal peduncle was contained from 1.4 to 1.8 times in its length. There are large cycloid scales on the body except on the head and fins (Tables 13, 17, 21 and 25).

Table 24: Recial Analysis of Mahseer, *Tor putitora* in Downstream of Ganga River–Regression Coefficient and Equation of Various Body Parts in Relation Total Length and Head Length

Body Parameters	*b–Value with Total Length (TL)*	
	Regression Coefficient	*Regression Equation*
Standard Length (SL)	0.81164	Y = 0.8116 X – 0.2944
Head Length(HL)	0.19735	Y = 0.1973 X + 0.0118
Max. Width (MW)	0.01464	Y = 0.0146 X + 11.9836
Max. Girth (MG)	0.02084	Y = 0.0208 X + 5.4228
Avg. Height (Avg. H)	0.01657	Y = 0.0165 X + 5.5790
Length of Caudal fin (LCF)	0.02474	Y = 0.0247 X + 6.1002
Depth of Caudal fin (DCF)	0.01310	Y = 0.0131 X + 3.3246
Length of Dorsal fin (LDF)	0.01713	Y = 0.0171 X + 4.7386
Depth of Dorsal fin (DDF)	0.00718	Y = 0.0071 X + 3.2386
Length of Pectoral fin (LPF)	0.01348	Y = 0.0134 X + 4.4644
Length of Pelvic fin (LPel.F)	0.01510	Y = 0.0151 X + 3.9666
Length of Anal fin (LAF)	0.01876	Y = 0.0187 X + 3.7342
Snout Length (SnL)	0.00455	Y = 0.0045 X + 4.0570
Eye Diameter (ED)	0.00101	Y = 0.0010 X + 1.1660
Inter Orbital Distance (IOD)	0.11860	Y = 0.1186 X – 0.8024
Length of Rostral barbel (LRB)	0.00274	Y = 0.0027 X + 1.7332
Length of Maxillary barbel (LMB)	0.00284	Y = 0.0028 X + 1.7598
Body Parameters	*b–Value with Head Length (HL)*	
	Regression Coefficient	*Regression Equation*
Snout Length (SnL)	0.22774	Y = 0.2277 X + 2.4522
Eye Diameter(ED)	0.14256	Y = 0.1425 X + 0.0999
Inter Orbital Distance (IOD)	0.59363	Y = 0.5936 X – 1.3523
Length of Rostral barbel (LRB)	0.06950	Y = 0.0695 X + 1.2884
Length of Maxillary barbel (LMB)	0.07196	Y = 0.0719 X + 1.3000

The mouth is small and the gape of the mouth was not wide. The lips are thick and fleshy and labial groove was continuous. There are two pairs of barbels of almost equal size and they are as long as the eye diameter. The colour of the fish is golden brown above and silvery white below. Adult fish is reddish green on the dorsal and light orange fading into silvery white at the mid-ventral line on the ventral surface and light pink at the sides. Fins are yellow.

Table 25: Meristic Characters of Mahseer, *Tor putitora* in Downstream of Ganga River

Characters	*Range*	*Mode*
Lateral line scale	24–27	27
Lateral line transverse scale	3½–4½	4½–3½
Pectoral fin rays	14–18	16
Dorsal fin lays	10–13 (2/9–4/9)	11 (4/9)
Caudal fin rays	18–22	19
Anal fin rays	10–13 (2/5–3/5)	13 (2/5–3/5)
Ventral fin rays	8–10	9

Fin formulas were observed both upstream and downstream sites for *Tor tor* as follows:

D11(2–3/9),P17,V9,A7,C19,L1,25 Ltr. 4½–4½ (for upstream)

D11(2–3/9),P17,V9,A7,C17,L1,23 Ltr. 4½–4½ (for downstream)

The fins are red in colour. The head is shorter than the depth of the body. The length of the head is contained 4.75 to 5.8 and the height of the body is 5½ to 6 times in the total length of the body. The diameter of the eye is contained 6.25 to 7.5 in the length of the head, in a moderately sized specimen. The lips are thick with an uninterrupted groove across the lower jaw. The maxillary pair of barbels are longer than the rostral.

Fin formulas of *Tor putitora* were calculated in upstream and downstream of Ganga river as follows:

D13(2–4/9),P17,V9,A7(2/5),C20,L1,27 Ltr. 4½–3½ (for upstream)

D11(2–4/9),P16,V9,A13(2–3/5),C19,L1,27 Ltr. 4½–3½ (for downstream)

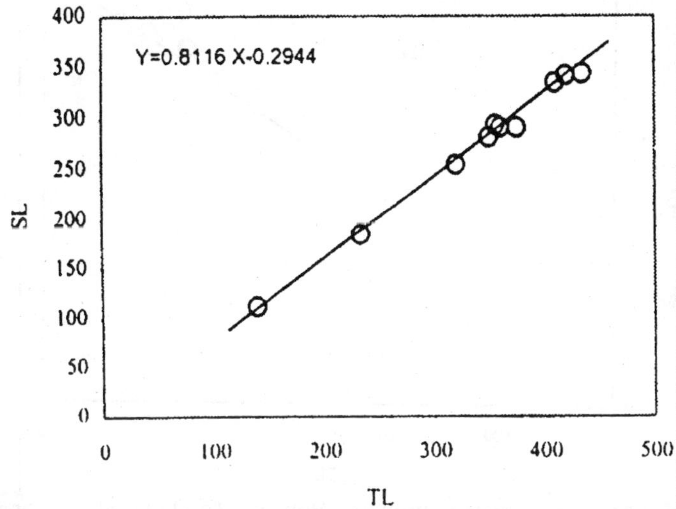

Figure 13(a): Linear Regression Between Total Length (TL) and Standard Length (SL) of *Tor putitora* in Downstream of Ganga River

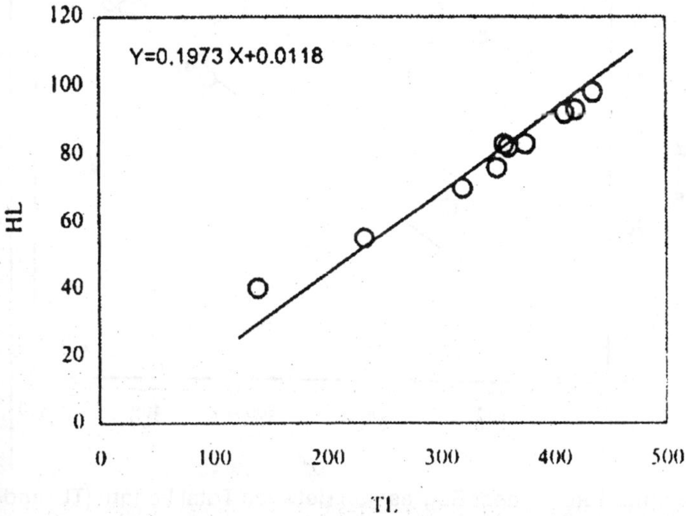

Figure 13(b): Linear Regression Between Total Length (TL) and Head Length (HL) of *Tor putitora* in Downstream of Ganga River

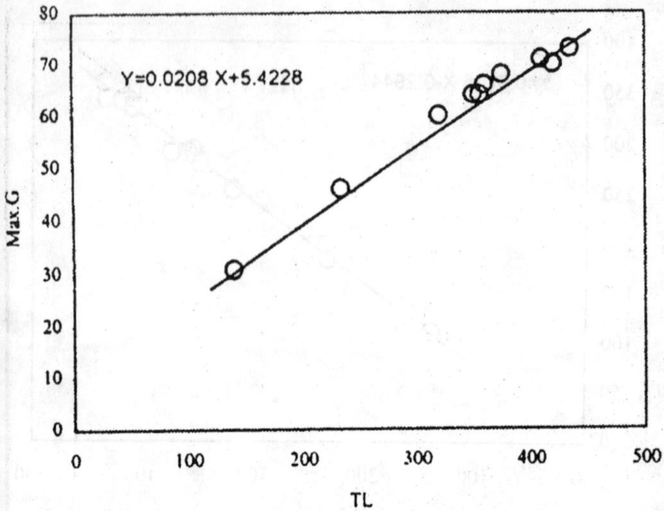

Figure 13(c): Linear Regression Between Total Length (TL) and Max. Girth (MG) of *Tor putitora* in Downstream of Ganga River

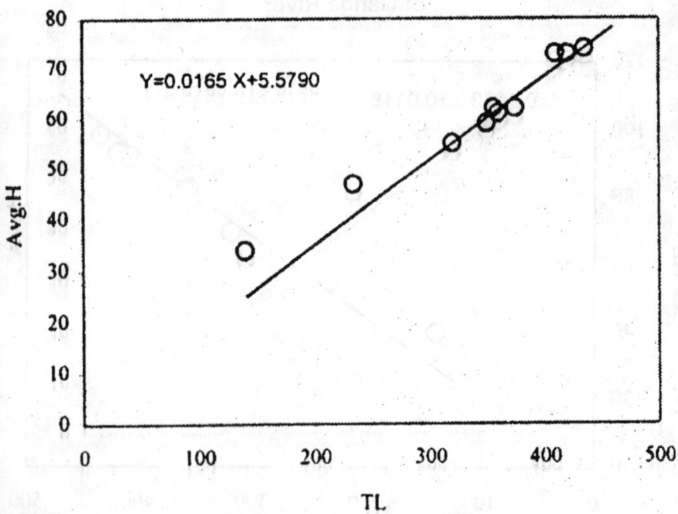

Figure 13(d): Linear Regression Between Total Length (TL) and Avg. H (Avg. H) of *Tor putitora* in Downstream of Ganga River

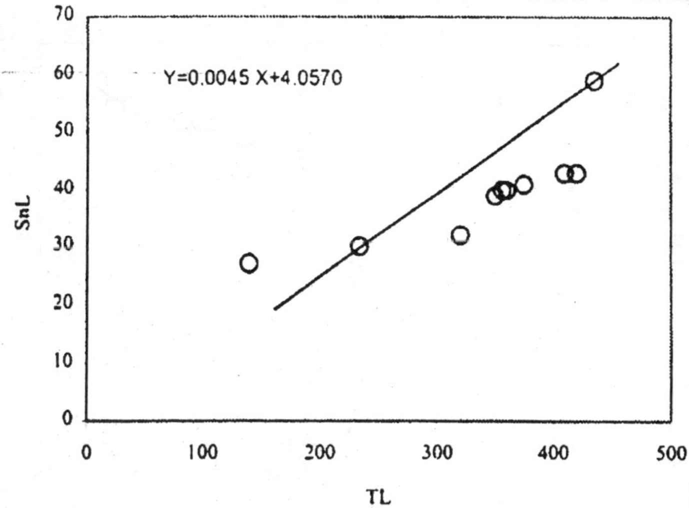

Figure 13(e): Linear Regression Between Total Length (TL) and Snout Length (SnL) of *Tor putitora* in Downstream of Ganga River

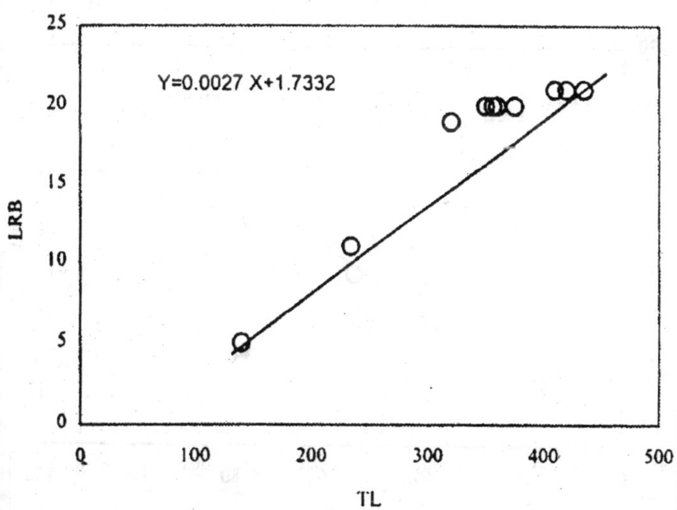

Figure 13(f): Linear Regression Between Total Length (TL) and Length of Rostral Barbel (LRB) of *Tor putitora* in Downstream of Ganga River

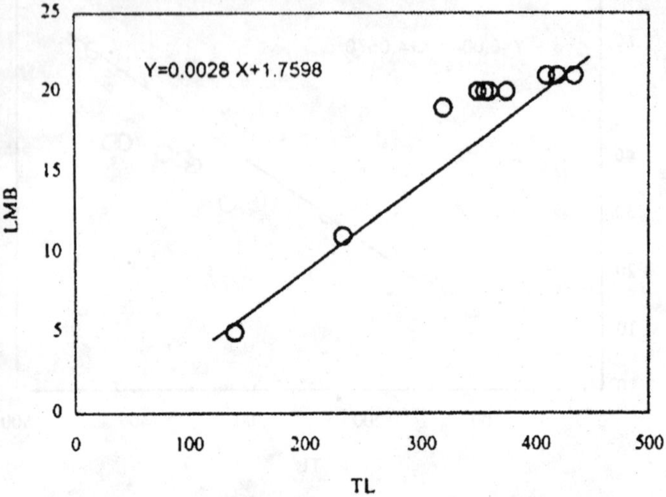

Figure 13(g): Linear Regression Between Total Length (TL) and Length of Maxillary Barbel (LMB) of *Tor putitora* in Downstream of Ganga River

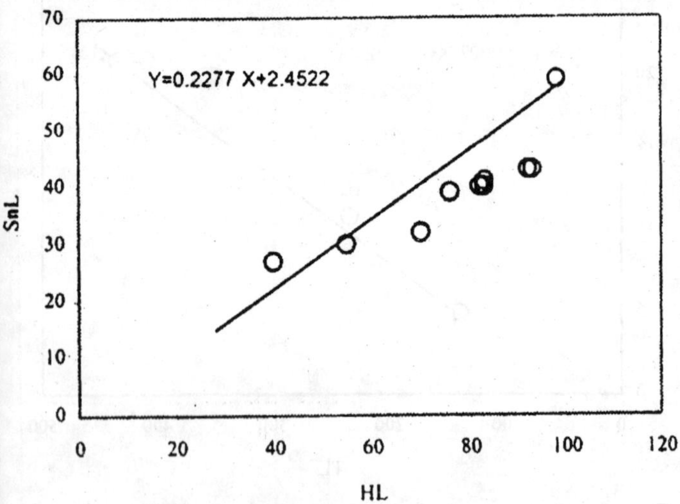

Figure 13(h): Linear Regression Between Head Length (HL) and Snout Length (SnL) of *Tor putitora* in Downstream of Ganga River

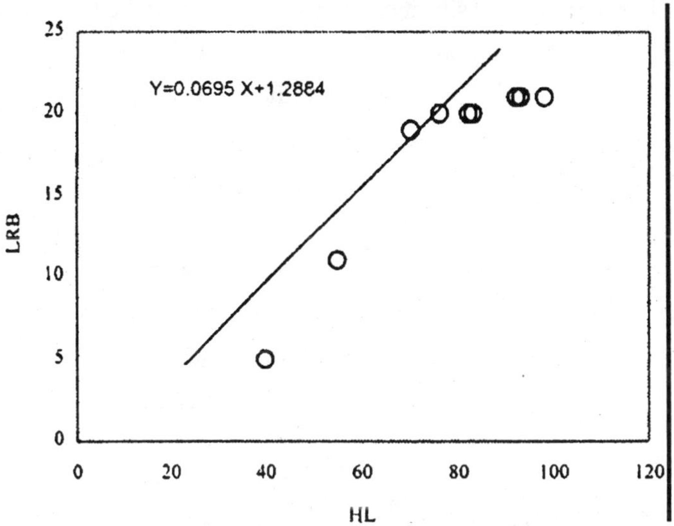

Figure 13(i): Linear Regression Between Head Length (HL) and Length of Rostral Barbel (LRB) of *Tor putitora* in Downstream of Ganga River

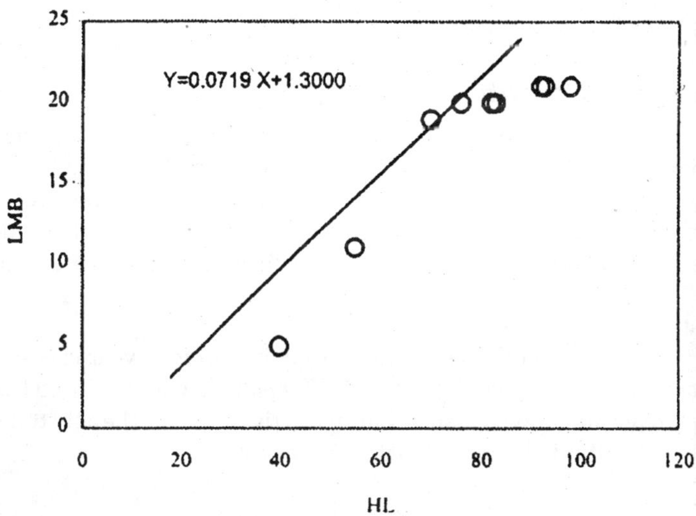

Figure 13(j): Linear Regression Between Head Length (HL) and Length of Maxillary Barbel (LMB) of *Tor putitora* in Downstream of Ganga River

The complete classification of mahseer has described as follows:

Kingdom:	Animalia
Phylum:	Chordata
Sub Phylum:	Vertebrata
Class:	Osteichthyes
Sub Class:	Actinopterygii
Division:	Haleostomi
Sub Division:	Teleostai
Infra Order:	Euteleostei
Super Order:	Ostariophysi
Series:	Otophysi
Order:	Cypriniformes
Family:	Cyprinidae
Genus:	*Tor*
Species:	*tor/putitora*

Age and Growth

Age and growth study of fish is essential for fishery management as also of great biological interest. Many methods of age determination in fishes are known, while in the present investigation the methods were used by interpretation of annulus formation laid down on hard parts, such as scale and operculum.

By Scales

In the present investigations, only one to four year classes of *Tor tor* and one to five year classes of *Tor putitora* were obtained from upstream and downstream of barrage in river Ganga. The age studied by the method of scales of mahseer is as follows:

One Year

The scales were examined by micro-projector method on a white screen, which clearly showed the circuli around the nucleus. The circuli were most prominent in the interior region of the scale. Therefore, the main observations for annuli were made in this region.

The posterior region of the scale also showed some annuli, which were at times used to avoid errors in age assessments. A narrow dark winter ring formed by one of the four methods forms each annulus. In between two annuli lies a wider light band of separated circuli constituting the summer ring. These winter and summer rings together from one year's growth of the scale. Only one annulus was present in the scale of this year class. The average length of the one year scale was observed 2.70 mm and the average total length of fish was recorded 110.00 mm in *Tor tor,* 1.95 mm average length of scale and 152.50 mm average total length were observed in *Tor putitora* (Table 26 and Figures 14a & 15a).

Table 26: Age Determination of Mahseer, *Tor tor* by Scales and Operculum in Upstream of Ganga river

Total Observed Length Range (mm)	*Average Length (mm)*	*Age by Scale*	*Age for Operculum*
160–240	203.33	2–2^{+++}	2
250–330	325.00	3–3^{+++}	3
350–440	380.00	4–4^{+++}	4

Two Years

There are two annuli indicating two years age of the fish. The average scale lengths were recorded 3.466 mm and 4.116 mm, while the average total length of fishes were observed as 215.00 mm and 261.50 mm in *Tor tor* and *Tor putitora* respectively. In the second year scale, the end of the first year's growth appeared as a clear streak in antero lateral part of the scale. This may be due to cessation of food intake in the last part of the year. Many interspersed small streaks were also seen, which represent false rings due to seasonal feeding fluctuations. The false rings may be avoided after continuous observations on a number of scales from the same fish. The age of the mahseer fishes were also confirmed by operculum methods.

Three Year

The annuli are seen as three dark marks with some streaks or crossing over of circuli. False streaks were also present due to spawning, which unlike the seasonal feeding fluctuation, streak only of first and second years have definite spawning marks. In the

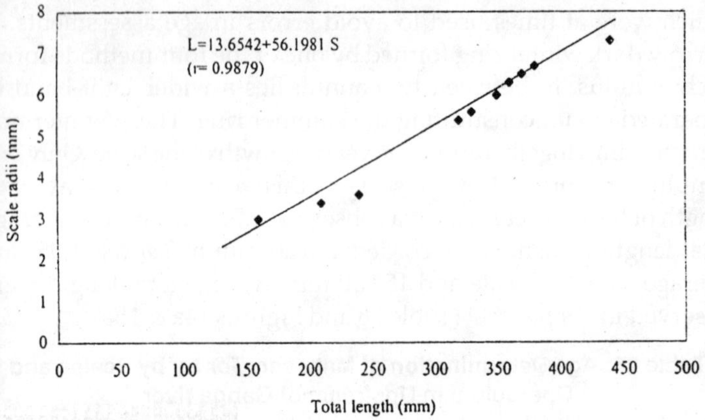

Figure 14(a): Fish Length Relationship with Scale Radii of *Tor tor* in Upstream of Ganga River

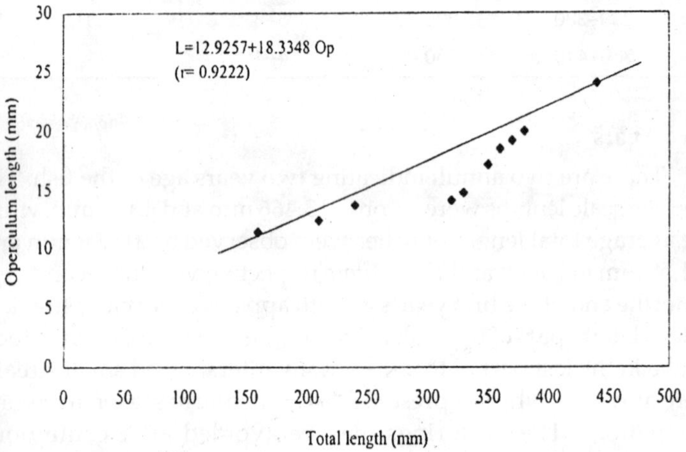

Figure 14(b): Fish Length Relationship with Operculum Length of *Tor tor* in Upstream of Ganga River

three year, the fishes have the average scale length were observed 5.650 mm and 5.577 mm and the average total length of fish were

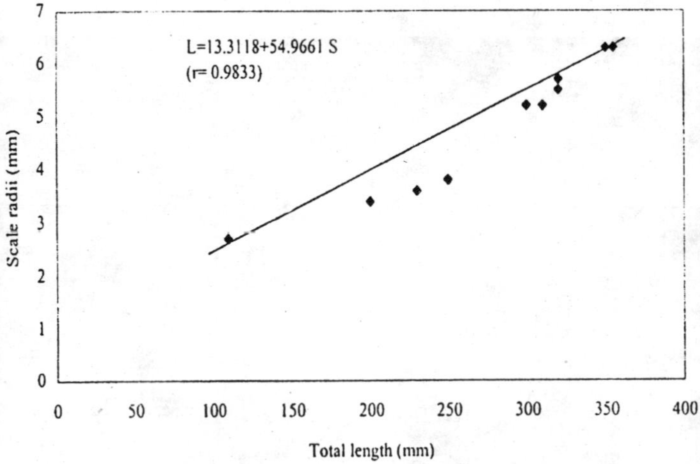

Figure 15(a): Fish Length Relationship with Scale Radii of *Tor tor* in Donstream of Ganga River

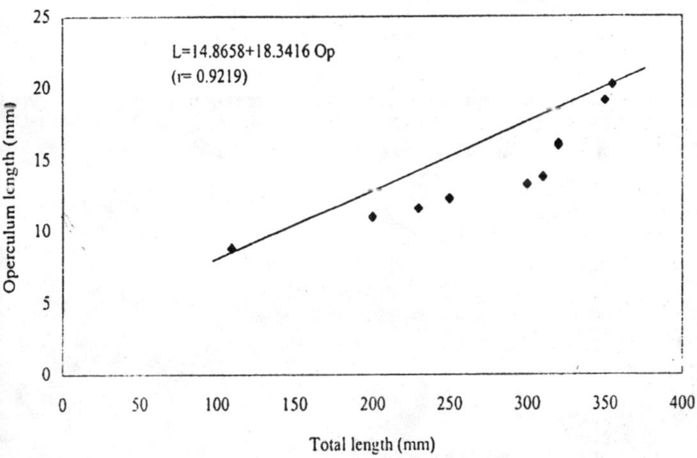

Figure 15(b): Fish Length Relationship with Operculum Length of *Tor tor* in Donstream of Ganga River

recorded 325.625 mm and 346.00 mm in *Tor tor* and *Tor putitora* respectively (Table 29, Figures 14a, 15a and Plates 8 & 9).

Plate 8: Growth Rings on the Cycloid Scale of Mahseer (*Tor putitora*)

Plate 9: Cycloid Scales of Different Age Groups of Mahseer (*Tor putitora*)

Table 27: Back Calculated Length of Mahseer, *Tor tor* by Scales and Operculum in Upstream of Ganga River

Age of Fish (in years)	Observed Total Length (mm)		Body Weight (gm)		Calculated Length by Scale (mm)		Calculated Length by Operculum (mm)	
	Range	Mean	Range	Mean	Range	Mean	Range	Mean
II–II+++	160–240	203.33	200–400	316.66	150.50–250.00	215.96	145.50–242.20	212.16
III–III+++	250–330	325.00	300–400	375.00	250.50–328.40	322.74	245.00–323.00	315.47
IV–IV+++	350–440	380.00	350–800	505.00	350.80–441.00	385.68	345.55–485.00	376.31

Table 28: Condition Factors (K-factor) of Mahseer, *Tor tor* in Upstream if Ganga River

Seasons	*Total Length of Fish Range (mm)*	*Total Fish Weight Range (gm)*	*K-factor Range*	*Average K-factor*
Summer	160–330	200–400	1.1130–4.8828	2.8984
Rainy	240–370	300–500	0.9871–2.1701	1.4084
Winter	350–440	350–500	0.7501–0.9391	0.8687

Table 29: Age Determination of Mahseer, *Tor tor* by Scales and Operculum in Downstream of Ganga river

Total Observed Length Range (mm)	*Average Length (mm)*	*Age by Scale*	*Age for Operculum*
080–140	110.00	$1\text{–}1^{+++}$	1
175–250	226.66	$2\text{–}2^{+++}$	2
290–360	325.83	$3\text{–}3^{+++}$	3

Four Years

Four clear annuli are present in the scale of fourth year class, with some false rings indicating fluctuations. Here also four dark narrow winter rings, followed light wide summer rings marking each year were observed. False rings or streaks and cutting over of circuli due to spawning and feeding fluctuations were also seen, as in the annuli of the third year class. There are four annuli and thus the fish is four year old. The average scale length were recorded 6.56 mm and 6.938 mm and the average total length of fishes were observed 380.00 mm and 424.888 mm of *Tor tor* and *Tor putitora* in four years age classes have observed (Tables 32 and Figures 16a & 17a).

Five Years

There are five annuli having many intermediary false rings after the third annulus, which appears to be spawning rings due to fluctuations in feeding during spawning period. The average scale length was found 8.15 mm and the total length was recorded 510.00 mm in *Tor putitora*, no any *Tor tor* was found is five year age class (Table 35).

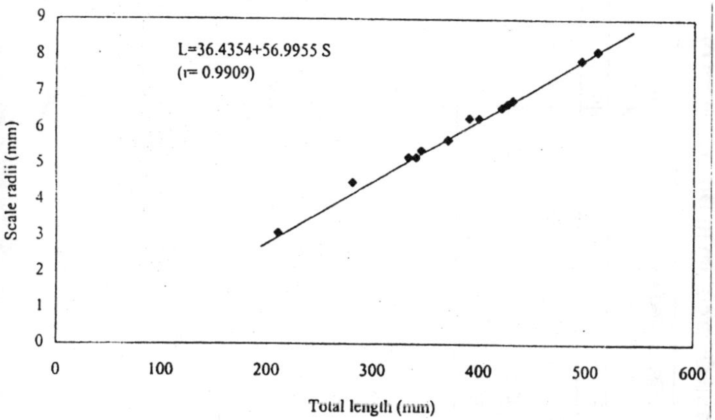

Figure 16(a): Fish Length Relationship with Scale Radii of *Tor putitora* in Upstream of Ganga River

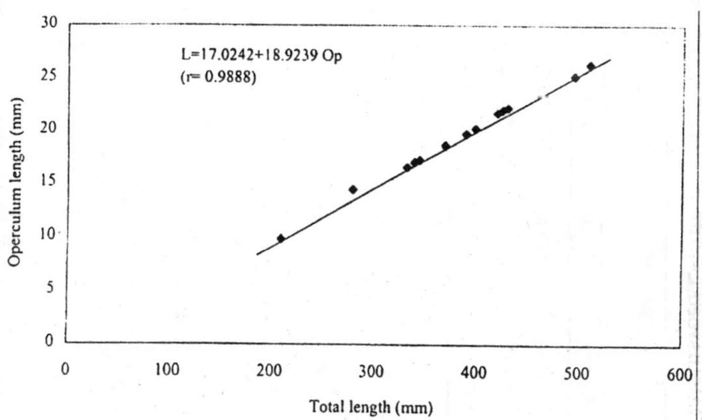

Figure 16(b): Fish Length Relationship with Operculum Length of *Tor putitora* in Upstream of Ganga River

Table 30: Back Calculated Length of Mahseer, *Tor tor* by Scales and Operculum in Downstream of Ganga River

Age of Fish (in years)	*Observed Total Length (mm)*		*Body Weight (gm)*		*Calculated Length by Scale (mm)*		*Calculated Length by Operculum (mm)*	
	Range	*Mean*	*Range*	*Mean*	*Range*	*Mean*	*Range*	*Mean*
I–I^{+++}	80–140	110.00	50–150	75.00	86.50–145.00	117.74	85.00–145.00	115.74
II–II^{+++}	175–250	226.66	190–300	246.66	180.00–270.00	236.91	180.00–262.50	233.73
III–III^{+++}	290–360	325.83	250–500	325.00	298.75–376.50	339.43	295.50–372.60	334.92

Table 31: Condition Factor (K-factor) of Mahseer, *Tor tor* in Downstream of Ganga river

Seasons	*Total Length of Fish Range (mm)*	*Total Fish Weight Range (gm)*	*K-factor Range*	*Average K-factor*
Summer	110–355	075–500	0.7629–5.6348	2.1725
Rainy	200–300	190–250	0.9259–2.3750	1.7852
Winter	250–350	300–300	0.6997–1.9200	1.1784

Table 32: Age Determination of Mahseer, *Tor putitora* by Scale and Opreculum Bone in Upstream of Ganga River

Total Observed Length Range (mm)	*Average Length (mm)*	*Age by Scale*	*Age for Operculum*
191–278	210.00	2–2+++	2
279–380	333.66	3–3+++	3
388–498	426.50	4–4+++	4
499–530	510.00	5–5+++	5

By Operculum

The opercular series have contained a number of investing bones such as opercular, pre-opercular, sub-opercular and inter-opercular. It has a slight concavity on the inner surface and was partly overlapped by the pre-opercular bone. The dorsal edge of the opercular is firmly attached to the pterotic of the skull by means of ligament and the ventral border was bounded by the sub-opercular. A facet for articulation with hyomandibular was present on the inner side of the opercular bone. The opercular bone of the fish an important hard part of the. body to assess the age of fish and to confirm the age by scale method (Plate 10). Each annulus is formed by many thin approximated rings around the border part of the opercular bone.

One Year

In the first year fish has a more translucent flexible opercular bone when compared to older fish operculum. The opercular bone has many ridges or streaks over the broader parts, which are

Table 33: Back Calculated Length of Mahseer, *Tor putitora* by Scale and Operculum in Upstream of Ganga River

Age of Fish (in years)	*Observed Total Length (mm)*		*Body Weight (gm)*		*Calculated Length by Scale (mm)*		*Calculated Length by Operculum (mm)*	
	Range	*Mean*	*Range*	*Mean*	*Range*	*Mean*	*Range*	*Mean*
II–II+++	191–278	210.00	110–200	150.00	190.00–240.50	213.12	195.50–275.00	223.29
III–III+++	279–380	333.66	201–500	352.00	270.00–365.00	326.54	289.00–369.50	333.80
IV–IV+++	388–498	426.50	500–1000	700.00	395.00–503.80	428.74	389.50–494.00	430.18
V–V+++	499–530	510.00	1100–1265	1250.00	504.00–550.00	526.59	495.00–532.30	522.29

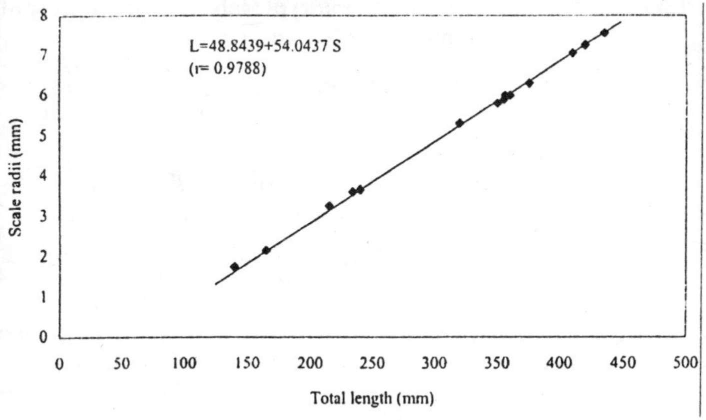

Figure 17(a): Fish Length Relationship with Scale Radii of *Tor putitora* in Downstream of Ganga River

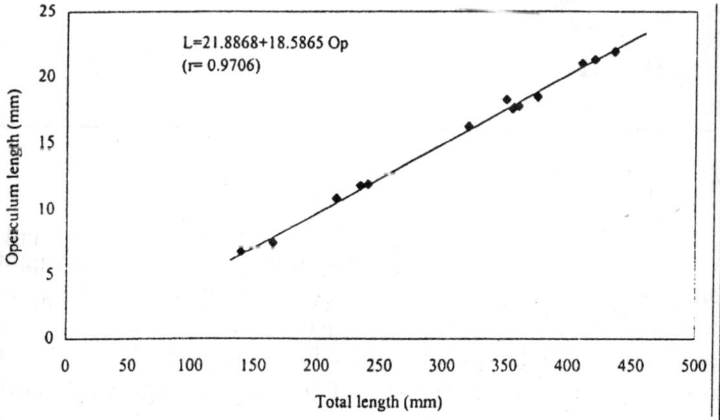

Figure 17(b): Fish Length Relationship with Operculum Length of *Tor putitora* in Downstream of Ganga River

compressed into a clear annulus near margin of the opercular bone indicating a one-year mark. As the operculum has only one annular mark, the fish has attained one year of age. The average width of opercular of one year class was observed 8.80 mm, in *Tor tor* and 7.15 mm in *Tor putitora*. This was also confirmed by the scale reading for one year age class.

Table 34: Condition Factor (K-factor) of Mahseer, *Tor putitora* in Upstream of Ganga River

Seasons	*Total Length of Fish Range (mm)*	*Total Fish Weight Range (gm)*	*K-factor Range*	*Average K-factor*
Summer	210–510	150–1250	0.9271–1.6196	1.1554
Rainy	345–425	450–750	0.9118–1.0958	1.0017
Winter	333–430	300–700	0.7632–0.8804	0.8107

Table 35: Age Determination of Mahseer, *Tor putitora* by Scales and Operculum in Downstream of Ganga River

Total Observed Length Range (mm)	*Average Length (mm)*	*Age by Scale*	*Age for Operculum*
130–180	152.50	$1\text{–}1^{+++}$	1
214–350	271.80	$2\text{–}2^{+++}$	2
351–399	361.50	$3\text{–}3^{+++}$	3
400–450	421.66	$4\text{–}4^{+++}$	4

Two Years

In two year class, the opercular is also more flexible than in older specimens, having two clear dark annuli between many lighter streaks in the border part of the operculum. Slight false rings between the annuli may be due to feeding fluctuations. Thus, the fish attained two years of age. The average opercular width were recorded 12.066 mm and 13.158 mm in *Tor tor* and *Tor putitora* respectively (Table 26).

Three Years

The opercular bone of three year class has three annular marks, each of which is a prominent ring at the broadest part of the operculum. Some false streaks were also present due to feeding fluctuations and spawning of fish. The average opercular width were recorded 15.925 mm in *Tor tor* and 17.255 mm in *Tor putitora* respectively (Table 29 and Figures 14b & 15b).

Plate 10: Growth Rings on the Opercular Bones of Mahseer (*Tor putitora*)

Four Years

The opercular bone has four annuli indicating the age as four years. However, false streaks appear between third and fourth years of age due to fluctuations in spawning stages. Other false streaks are also visible, which may be due to feeding fluctuations. The average opercular width were recorded 19.780 mm and 21.688 mm in *Tor tor* and *Tor putitora* respectively (Table 32 and Figures 16b & 17b). The age was also confirmed by the another method .

Five Years

The opercular bone has five annuli indicating the fish to be five years old. The false streaks seen in third and fourth year classes were also marked. However, the false rings and streaks were more in this year class indicating more fluctuations in feeding than in the four year old. The average opercular width was found 26.30 mm in *Tor putitora.* No any five years old *Tor tor* was found in upstream and downstream of barrage in Ganga river (Table 35).

Back Calculation

Back calculation by scales and opercular bones are an important aspect to determine the total length of fish of younger age classes, using only scale annuli and opercular width as determinants. By establishing a formula for relationships between radius of scale and total length of fish, and between opercular width and total length of the fish, a straight line has been established for scale and opercular.

The relationship between scale radius and fish length and opercular width and fish length have been calculated and expressed as for *Tor tor:*

$$L = 13.6542 + 56.1981\ S$$

(by scale)

$$L = 12.9257 + 18.3348\ Op$$

(by opercular bone)

The equations for upstream of barrage in Ganga river for *Tor tor* and

$$L = 13.3118+54.9661\ S$$

(by scale)

$$L= 14.8658 + 18.3416\ Op$$

(by opercular bone)

Equations for downstream of barrage in Ganga river for *Tor tor*.

The relationship between scale radius and total length and opercular width and total length of *Tor putitora* for upstream and downstream of barrage have been calculated and expressed as:

$$L = 36.4354 + 56.9955\ S$$

(by scale)

$$L= 17.0242 + 18.9239\ Op$$

(by opercular bone)

Above equations for upstream of barrage in Ganga river and

$$L= 48.8439 + 54.0437\ S$$

(by scale)

$$L= 21.8868 + 18.5865\ Op$$

(by opercular bone)

Table 36: Back Calculated Length of Mahseer, *Tor putitora* by Scales and Operculum in Downstream of Ganga River

Age of Fish (in years)	*Observed Total Length (mm)*		*Body Weight (gm)*		*Calculated Length by Scale (mm)*		*Calculated Length by Operculum (mm)*	
	Range	*Mean*	*Range*	*Mean*	*Range*	*Mean*	*Range*	*Mean*
I–I^{+++}	130–180	152.50	45–80	62.50	125.50–175.00	158.27	110.00–180.00	156.63
II–II^{+++}	214–350	271.80	89–331	180.00	201.00–349.00	274.74	221.00–350.00	271.31
III–III^{+++}	351–399	361.50	348–502	423.75	351.00–421.00	359.59	349.00–410.50	354.11
IV–IV^{+++}	400–450	421.66	595–720	643.33	421.50–470.50	427.14	411.00–465.00	419.63

Equations for downstream of barrage in Ganga river. From the above formulas, the calculated length for each year class were found significantly close to the observed total length as shown in the (Tables 27, 30, 33 & 36 and Figures 14-17).

Condition Factor (K-factor)

The condition factor (K-factor) is a well recognized aspect in fishery biology. In the present study, condition factor for three season summer, rainy and winter for two year period were made as follows:

Condition Factor for *Tor tor*

The average value of condition factor (K-factor) for both sites upstream and downstream varied from 0.8687 to 2.8984 and 1.1784 to 2.1725 for the two years in all three seasons. The highest K-factor was recorded in the summer season of both the years as 2.8984 for upstream and 2.1725 for downstream of barrage. The lowest values of K-factor 0.8687 were observed for upstream in the winter season and recorded 1.1784 for downstream. The highest values 2.8984 and 2.1725 of condition factor were obtained during breeding season (April to September) in upstream an downstream respectively, when slight monthly or seasonal variations have occurred due to the repeated spawning habits of mahseer; and also during pre and post spawning periods due to voracious feeding by the mahseer fishes. The K-factor was recorded minimum (0.8687 and 1.1784) during cold months of winter (Tables 28 & 31 and Figures 18a&b).

Table 37: Condition Factor (K-factor) of Mahseer, *Tor putitora* in Downstream of Ganga River

Seasons	*Total Length of Fish Range (mm)*	*Total Fish Weight Range (gm)*	*K-factor Range*	*Average K-factor*
Summer	165–356	075–450	0.7696–1.6695	0.9998
Rainy	140–420	050–610	0.8233–1.8221	1.0691
Winter	240–435	130–720	0.7629–0.9403	0.8621

Condition Factor (K-factor) for *Tor putitora*

The condition factor (K-factor) was calculated, which varied from winter to summer 0.8107 to 1.4554 in upstream and winter to rainy 0.8621 to 1.0691 in downstream of barrage. The highest

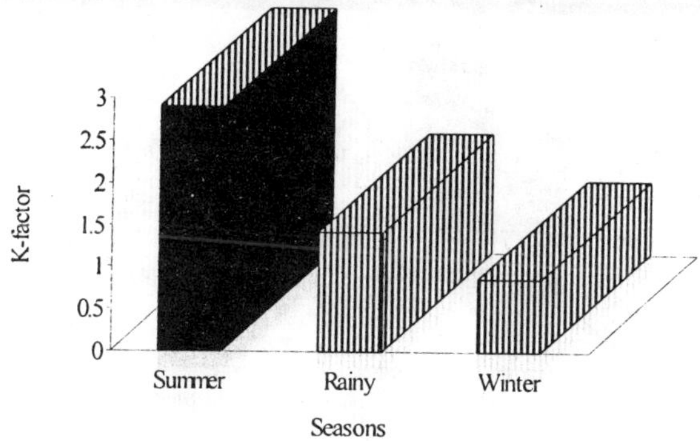

Figure 18(a): Condition Factor (K-factor) of *Tor tor* in Upstream of Ganga River

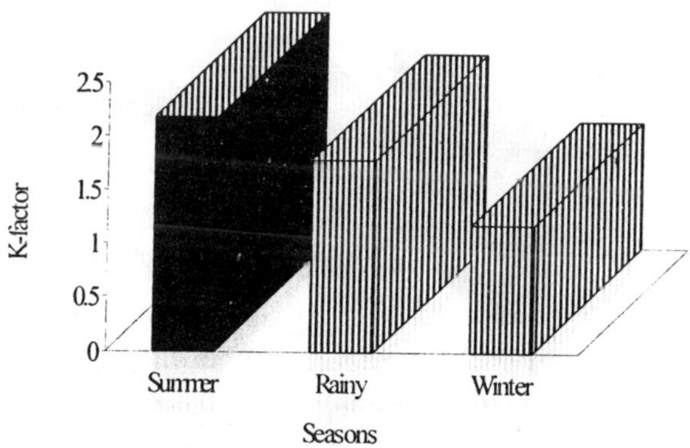

Figure 18(b): Condition Factor (K-factor) of *Tor tor* in Downstream of Ganga River

condition factor (1.4554 and 1.0017) were obtained in summer and rainy season in upstream and (1.0691 and 0.9998) were obtained in rainy and summer season in downstream in both the years. The highest values were obtained during breeding season in both the years during pre and post spawning periods (Tables 34, 37 and Figures 19a&b).

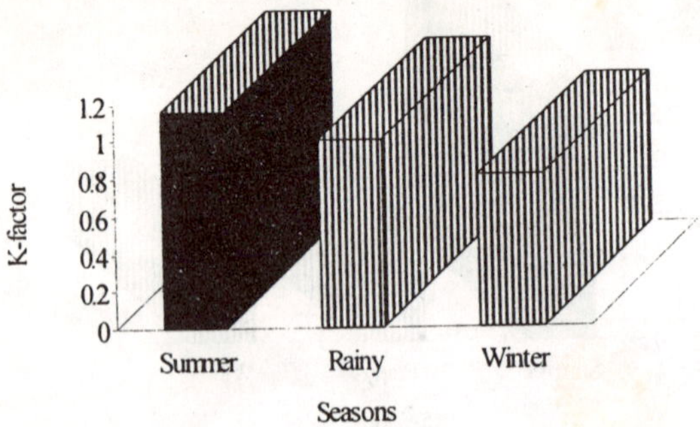

Figure 19(a): Condition Factor (K-factor) of *Tor putitora* in Upstream of Ganga River

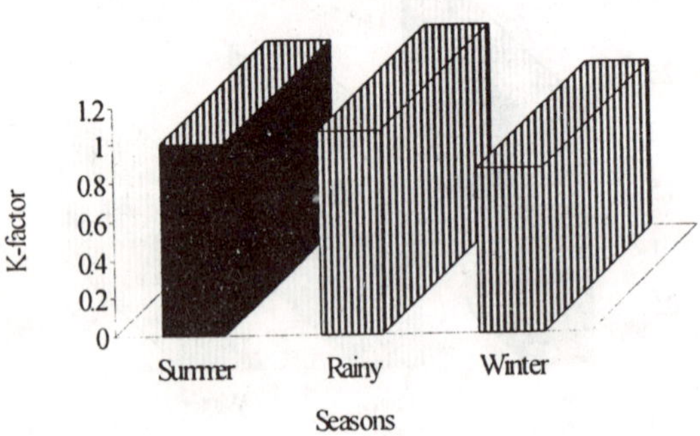

Figure 19(b): Condition Factor (K-factor) of *Tor putitora* in Downstream of Ganga River

Food and Feeding Habits

The size range of mahseer examined from 310.00 mm to 440.00 mm in total length of *Tor tor* and 234.00 mm to 480.00 in total length of *Tor putitora* in upstream and downstream of barrage in Ganga river at Rishikesh.

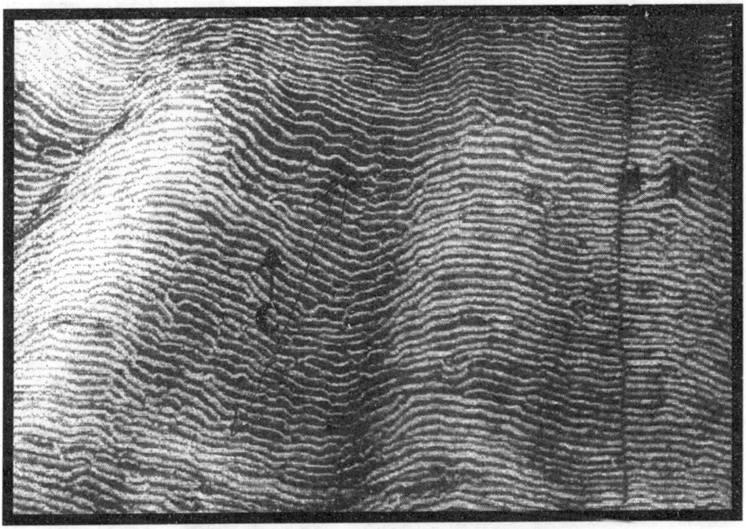

Plate 11: Scanning Electron Micrograph of the Scale of Mahseer (*Tor putitora*) Showing Details of Circuli and Radius

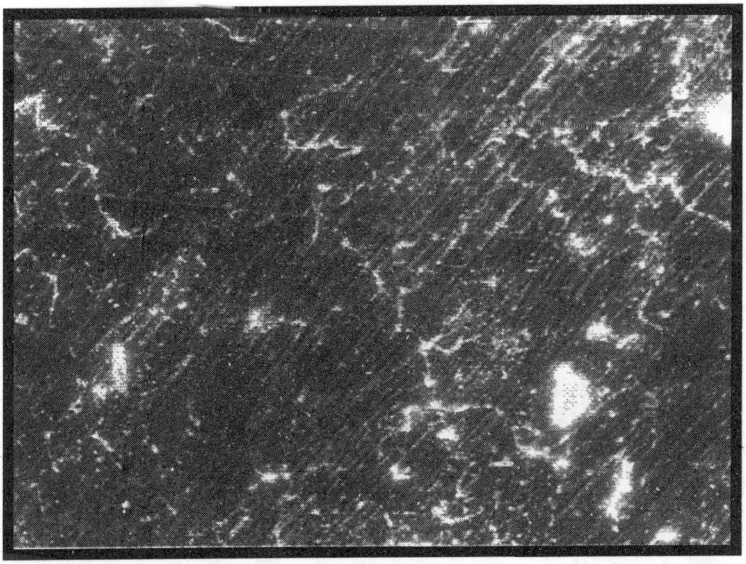

Plate 12: Scanning Electron Micrograph of Annular Zone on the Scale of Mahseer, (*Tor putitora*) Showing the Breaking on the Circuli

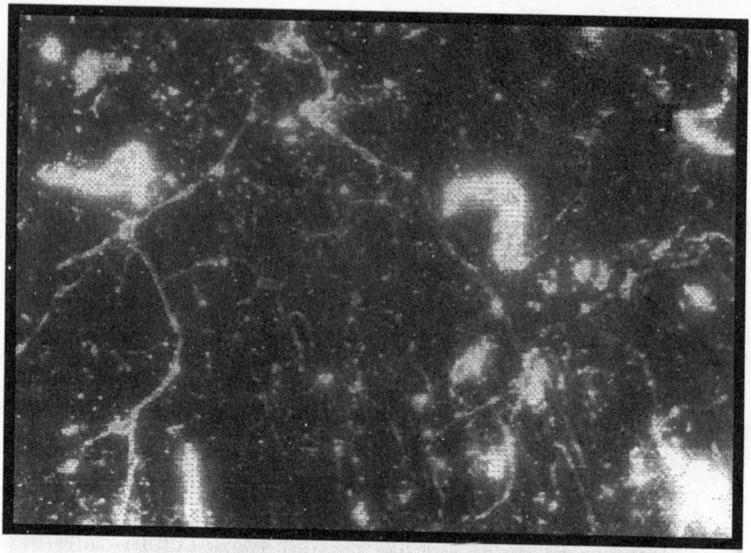

Plate 13: Scanning Electron Micrograph of the Outer Margin of the Scale of Mahseer, (*Tor putitora*) with Teeth Like Structures on the Circuli

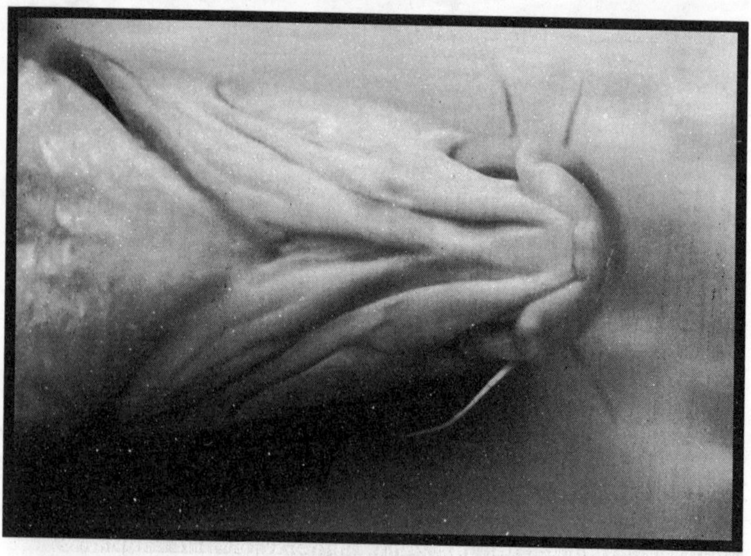

Plate 14: Ventral Surface of Head and Nature of Lower Lips and Jaws of Mahseer (*Tor putitora*)

Tor tor

Gross Morphology of Alimentary Canal

The buccopharynx open by a narrow cresentric mouth at the end of the snout, which is bounded by soft fleshy upper and lower lips. The mouth is followed by a soft toothless buccal cavity, which opens into an anterior wider region and a posterior narrow region provided with a callous pad. The tongue is not clearly differentiated in this fish. Many pharyngeal homodont teeth are embedded in the soft tissue of the ventral wall. These inferior pharyngeal teeth work against the dorsal callous pad thus work as a masicatory apparatus for crushing the hard part of food material. These organs are collectively known as buccopharnx and lead through oesophagus into the intestinal bulb.

The oesophagus is short, muscular and leads straight into the thin walled intestinal bulb, which is a swollen part of the intestine for food storage. A stomach is absent in *Tor tor* as evidenced by the opening of bile duct from the gall bladder into an instestinal bulb. The gall bladder opens in the anterior one third of the intestinal bulb. The intestinal bulb becomes narrow posteriorly and leads into the intestine proper, which is terminated by the rectum.

Seasonal Qualitative Fluctuations in Fish Food

Seasonal or monthly fluctuations in the food of fishes are well known. This is not only varies in different fish species but also in the same species due to availability of food organism in a particular water body, which the fish inhabits. Therefore, in the present study the fluctuations of different food items have been studied.

The food items identified are given in Tables 38–40, showing seasonal fluctuations of food items for *Tor tor*. Different aquatic weeds and algae were observed throughout all seasons during the period of investigations indicating plant food items as preferred food. Diatoms were also present in food items in all seasons. In animal food, different types of crustaceans and insects were always dominated showing there to be preferred in animal food items. Nematodes, rotifers, protozoans and some fish parts were also present in food items. The quantity of these items were recorded lower than crustaceans and insects. Some hairs, threads, sand particles and decaying organic matter etc. were also found in the gut of the fish, which shows *Tor tor* also feeds at bottom and side waters of river.

Table 38: Qualitative Assessment of Food Items in Gut of Mahseers, *Tor tor* and *Tor putitora* in Upstream of Ganga River

Sl. No.	*Food Items*	*Tor tor*			*Tor putitora*		
		Summer	*Rainy*	*Winter*	*Summer*	*Rainy*	*Winter*
	Macrophytes						
1.	*Vallisneria* sp.	++	++	+	+	–	++
2.	*Ceratophyllum* sp.	++	++	++	++	+	++
3.	*Hydrilla* sp.	+	+	++	+	–	++
4.	*Polygonium* sp.	–	+	++	–	+	+
5.	*Potamogeton* sp.	+	+	++	–	+	+
	Algae						
1.	*Spirogyra* sp.	++	++	++	++	+	++
2.	*Ulothrix* sp.	+	–	+	+	+	++
3.	*Zygnema* sp.	+	+	–	+	+	–
4.	*Pediastrum duplex*	++	–	++	++	+	++
5.	*Closterium* sp.	–	+	+	–	–	++
6.	*Chlorococcum* sp.	–	–	++	–	–	+
7.	*Chara* sp.	+	–	+	+	–	+
	Diatoms						
1.	*Nitzchia* sp.	+	+	+	+	++	+
2.	*Synedra* sp.	++	+	–	+	+	–
3.	*Diatoma* sp.	+	+	–	+	–	–
4.	*Navicula* sp.	+	++	+	+	++	+
	Insecta						
1.	Diptera larvae	+	+	–	++	++	+
2.	Hemiptera larvae	+	+.	–	+	–	+
3.	Chironomus larvae	+	–	+	+	+	+
4.	Neuroptera larvae	+	–	+	+	+	+
	Crustacea						
1.	*Daphnia* spp.	+	+	–	+	++	+
2.	*Cyclops*	–	–	+	++	++	+
	Nematods	–	–	–	+	+	+
	Protozoans	–	–	+	–	–	–
	Molluscs	–	–	+	+	+	+
	Fish parts	–	–	–	+	+	–

Table 39: Qualitative Assessment of Food Items in Gut of Mahseers, *Tor tor* and *Tor putitora* in Downstream of Ganga River

Sl. No.	*Food Items*	*Tor tor*			*Tor putitora*		
		Summer	*Rainy*	*Winter*	*Summer*	*Rainy*	*Winter*
	Macrophytes						
1.	*Hydrilla* sp.	+	++	−	+	+	++
2.	*Ceratophyllum* sp.	+	+	++	−	+	+
3.	*Vellisneria* sp.	+	++	++	−	−	+
	Algae						
1.	*Spirogyra* sp.	+	++	++	+	+	+
2.	*Ulothrix* sp.	−	+	+	−	+	+
3.	*Chlorococcum* sp.	−	+	+	−	++	+
4.	*Zygnema* sp.	+	+	−	−	+	−
	Diatoms						
1.	*Nitzchia* sp.	+	+	++	+	++	+
2.	*Synedra* sp.	++	+	++	−	+	−
3.	*Navicula* sp.	+	+	−	+	++	+
4.	*Diatoma* sp.	+	+	−	−	+	+
	Insecta						
1.	Diptera larvae	+	++	++	+	+	++
2.	Hemiptera larvae	+	+	++	−	−	+
3.	Chironomus larvae	+	−	++	−	I	++
4.	Scales of Lapidoptera	+	−	++	+	++	++
5.	Neuroptera larvae	+	−	−	+	+	−
6.	Coleoptera larvae	+	++	+	+	+	+
7.	Ephemeroptera nymphs	+	−	−	+	+	+
	Crustacea						
1.	*Daphnia* spp.	+	+	+	++	++	+
2.	*Cyclops*	+	+	++	++	++	++
3.	Cypris spp.	−	−	++	+	++	++
	Rotifers	−	+	+	−	−	+
	Protozoans	−	−	+	−	+	+
	Annelids	−	+	+	−	+	+
	Nematods	−	+	+	+	−	+
	Molluscs	−	+	+	+	−	+
	Fish parts	−	+	−	+	+	+

Table 40: Qualitative Assessment of Food Items in Gut of Mahseers, *Tor tor* and *Tor putitora* in Saptsarovar in Ganga River

Sl. No.	*Food Items*	*Tor tor*			*Tor putitora*		
		Summer	*Rainy*	*Winter*	*Summer*	*Rainy*	*Winter*
	Macrophytes						
1.	*Hydrilla* sp.	+	++	++	++	+	++
2.	*Potamogeton* sp.	+	+	++	+	+	++
3.	*Ceratophyllum* sp.	++	+	++	+	+	++
4.	*Polygonium* sp.	–	+	++	–	+	+
5.	*Vellisneria* sp.	++	++	–	–	+	+
	Algae						
1.	*Spirogyra* sp.	+	++	++	+	+	+
2.	*Ulothrix* sp.	+	–	+	+	+	+
3.	*Zygnema* sp.	+	+	–	+	+	–
4.	*Closterium* sp.	–	+	+	–	–	+
5.	*Chlorococcum* sp.	–	–	++	–	–	++
6.	*Chara* sp.	++	–	++	–	–	–
	Diatoms						
1.	*Nitzchia* sp.	+	+	++	+	++	++
2.	*Navicula* sp.	+	++	–	++	++	++
3.	*Diatoma* sp.	+	+	–	+	–	+
4.	*Fragilaria* sp.	++	–	–	+	–	–
5.	*Synedra* sp.	++	+	++	–	+	++
	Insecta						
1.	Diptera larvae	++	++	++	++	++	++
2.	Chironomus larvae	+	–	++	–	+	++
3.	Ephemeroptera nymphs	+	–	–	++	+	++
4.	Scales of Lapidoptera	+	–	++	+	+	++
5.	Hemiptera larvae	+	+	+	+	–	+
	Crustacea						
1.	*Daphnia* spp.	+	+	+	+	++	+
2.	*Cypris* spp.	–	–	+	++	++	++
3.	*Cyclops*	+	+	+	++	++	++
	Nematods	–	+	+	–	+	+
	Protozoans	–	–	+	–	–	+
	Rotifers	–	+	+	–	+	+
	Annelids	–	–	+	–	+	–
	Fish parts	–	+	–	+	+	–

Seasonal Quantitative Fluctuations in Food Items

The percentage of plant matter, animal matter and digested matter were estimated seasonally. The result for *Tor tor* in upstream, downstream of barrage and Saptsarovar are given as follows:

In the upstream, the highest value of plant matter was found in rainy season (80.73, 70.39 and 35.61 per cent), when more aquatic plants were available in the upstream of river Ganga. A high value of plant matter was also obtained in the winter season (23.63, 20.34 and 29.65 per cent) than in the summer season of the year, while the lower values were obtained in the summer (3.87, 2.04 and 2.27 per cent). Animal matter was found highest 12.18 per cent in rainy season and lowest was recorded 2.04 per cent in summer, whereas high percentage of digested matter was observed 88.26 per cent in summer and lowest recorded 2.10 per cent in rainy season. Sand and detritus were found in all three season (Table 41 and Figures 20a–c).

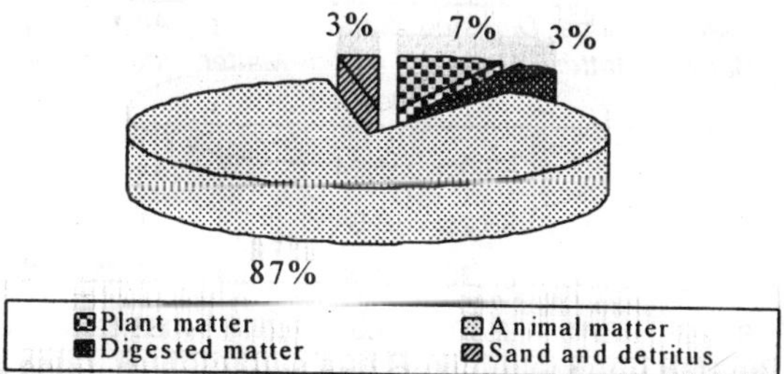

Figure 20(a): Quantitative and Qualitative Food Percentage of *Tor tor* During Summer Season in Upstream of Ganga River

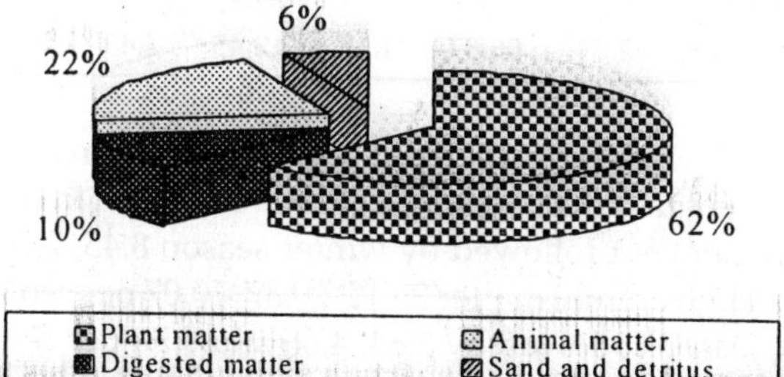

Figure 20(b): Quantitative and Qualitative Food Percentage of *Tor tor* During Rainy Season in Upstream of Ganga River

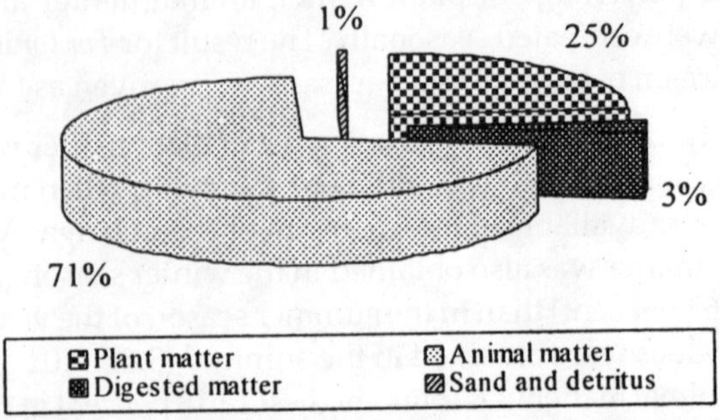

Figure 20(c): Quantitative and Qualitative Food Percentage of *Tor tor* During Winter Season in Upstream of Ganga River

Table 41: Seasonal Quantitative and Qualitative Food Percentage of Mahseers, *Tor tor* and *Tor putitora* in Upstream of Ganga River

Seasons	*Tor tor*				*Tor putitora*			
	Plant Matter	*Animal Matter*	*Digested Matter*	*Sand and Detritus*	*Plant Matter*	*Animal Matter*	*Digested Matter*	*Sand and Detritus*
Summer	8.99	3.87	85.90	1.24	7.37	16.62	74.51	1.49
	5.76	2.04	88.26	3.94	4.23	9.82	84.03	1.92
	7.03	2.27	87.70	3.00	5.09	12.21	80.24	2.46
Rainy	80.73	12.18	2.10	4.99	27.52	63.69	3.65	5.14
	70.39	9.41	14.80	5.40	30.19	55.25	8.83	5.73
	35.61	9.22	47.72	7.45	14.33	32.24	45.36	8.07
Winter	23.63	5.49	70.63	0.25	12.25	9.67	77.27	0.81
	20.34	2.10	77.37	0.19	15.21	10.32	73.38	1.09
	29.65	2.51	66.67	1.17	12.45	7.43	79.33	0.79

In the downstream of barrage, plant matter was recorded highest 59.59 per cent in rainy season and observed lowest in summer 7.91, 2.00, 2.48 per cent followed by winter season 8.45, 7.59 and 17.47 per cent. High animal matter 26.12, 20.37, 10.07 per cent was found in rainy season followed by 5.64, 4.68 and 8.36 per cent in winter and lowest value observed in summer season 4.72, 2.20 and 1.85 per cent. Digested matter 82.95, 89.66 and 90.16 per cent were found in summer and lowest quantity were found in rainy season 6.76, 27.78 and 44.72 per cent, sand and detritus was found in all the season, whereas the quantities of sand and detritus were highest 11.36 per cent in rainy season and recorded lowest quantity 0.57 per cent in winter season (Table 42 and Figures 21a–c).

Table 42: Seasonal Quantitative and Qualitative Food Percentage of Mahseers, *Tor tor* and *Tor putitora* in Downstream of Ganga River

Seasons	*Tor tor*				*Tor putitora*			
	Plant Matter	*Animal Matter*	*Digested Matter*	*Sand and Detritus*	*Plant Matter*	*Animal Matter*	*Digested Matter*	*Sand and Detritus*
Summer	7.91	4.72	82.95	4.42	1.11	13.76	78.98	6.15
	2.00	2.20	89.66	6.14	0.78	6.50	88.93	3.79
	2.48	1.85	90.15	5.52	2.10	7.65	84.45	5.80
Rainy	59.59	26.12	6.76	7.53	15.34	61.11	3.94	19.61
	42.64	20.37	27.78	9.21	7.97	36.42	40.25	15.36
	33.85	10.07	44.72	11.36	8.75	30.84	41.65	18.76
Winter	8.45	5.64	84.74	1.17	3.15	16.27	80.05	0.53
	7.59	4.68	87.16	0.57	5.89	24.99	68.38	0.74
	17.47	8.36	71.84	2.33	2.21	24.75	71.70	1.34

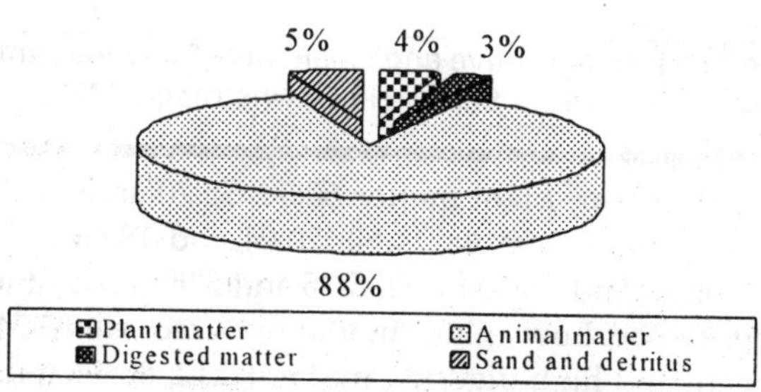

Figure 21(a): Quantitative and Qualitative Food Percentage of *Tor tor* During Summer Season in Downstream of Ganga River

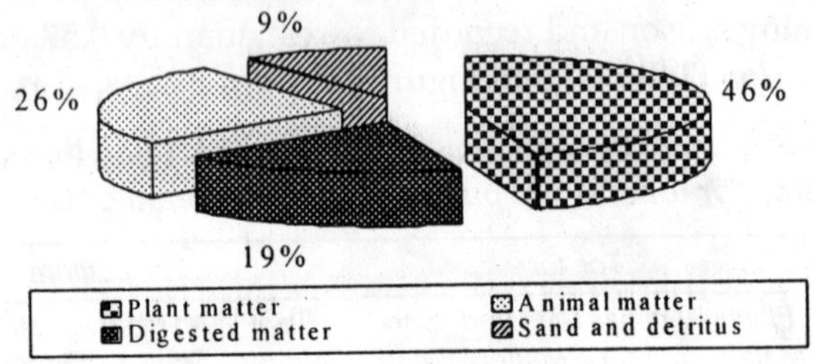

Figure 21(b): Quantitative and Qualitative Food Percentage of *Tor tor* During Rainy Season in Downstream of Ganga River

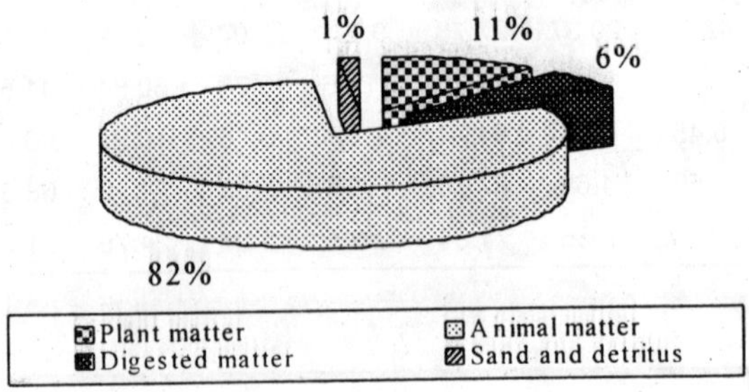

Figure 21(c): Quantitative and Qualitative Food Percentage of *Tor tor* During Winter Season in Downstream of Ganga River

At Saptsarovar, plant matter 60.30, 55.20 and 23.81 per cent was found in rainy season, 14.80, 20.04 and 18.66 per cent was found in winter and lowest 7.54, 5.35 and 5.90 per cent was found in summer season. Low animal matter were recorded 3.54, 2.66 and 1.26 per cent and high digested matter 82.12, 86.94 and 86.81 per cent was also found in summer season. Quantity of animal matter 26.27, 17.34 and 12.98 per cent was found in rainy season and lowest quantity of animal matter was recorded in winter season as 5.74, 3.80 and 4.86 per cent. Lowest digested matter was found 2.10, 16.71 and 52.83 in rainy season. Sand and detritus was found in all three season, the quantity was recorded maximum in rainy and minimum in winter season during the study period (Table 43 and Figures 22a–c).

Gastrosomatic Index (GSI)

The gastrosomatic index is a relation of weight of alimentary canal to weight of fish body, which may be considered as an index of variations in food items intake. Therefore, gastrosomatic index was studied in summer, rainy and winter season during the study period.

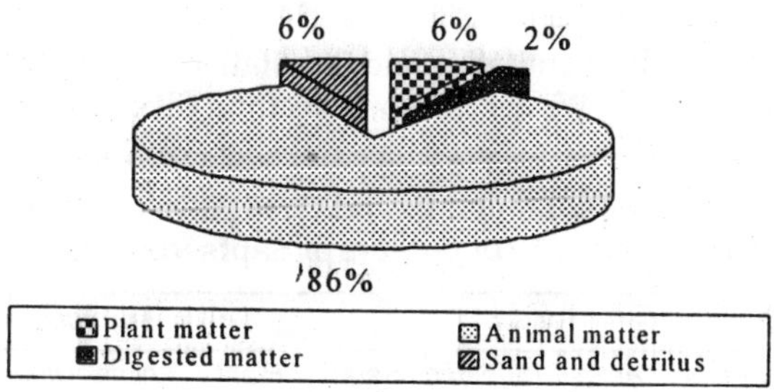

Figure 22(a): Quantitative and Qualitative Food Percentage of *Tor tor* During Summer Season in Ganga Riverat Saptsarovar

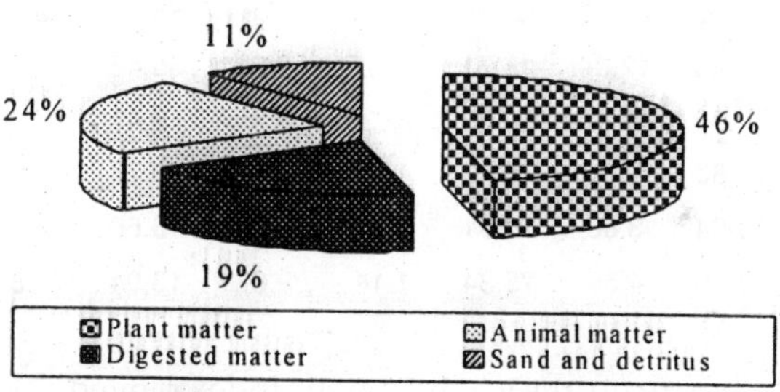

Figure 22(b): Quantitative and Qualitative Food Percentage of *Tor tor* During Rainy Season in Ganga River at Saptsarovar

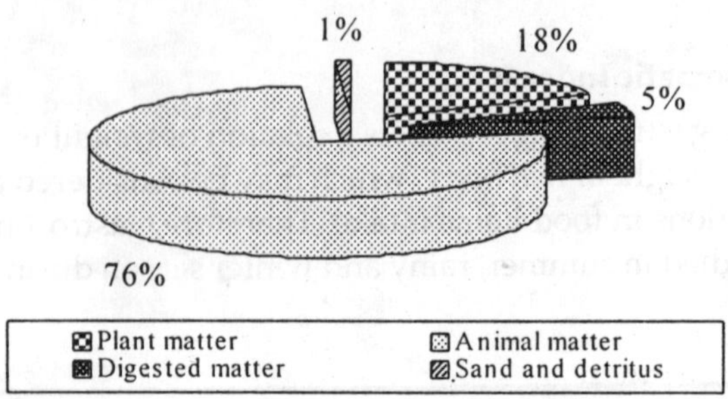

Figure 22(c): Quantitative and Qualitative Food Percentage of *Tor tor* During Winter Season in Ganga River at Saptsarovar

Table 43: Seasonal Quantitative and Qualitative Food Percentage of Mahseers, *Tor tor* and *Tor putitora* at Saptsarovar in Ganga River

Seasons	*Tor tor*				*Tor putitora*			
	Plant Matter	*Animal Matter*	*Digested Matter*	*Sand and Detritus*	*Plant Matter*	*Animal Matter*	*Digested Matter*	*Sand and Detritus*
Summer	7.54	3.54	82.12	6.80	1.35	9.11	80.57	8.97
	5.35	2.66	86.94	5.05	1.98	2.29	84.11	11.62
	5.90	1.26	86.81	6.03	1.00	6.23	85.20	7.57
Rainy	60.30	26.27	2.10	11.33	10.81	58.94	19.72	10.53
	55.20	17.34	16.71	10.75	6.93	55.48	24.26	13.33
	23.81	12.98	52.83	10.36	8.85	43.45	37.07	10.63
Winter	14.80	5.74	78.42	1.04	8.26	19.78	69.77	2.19
	20.04	3.80	74.91	1.25	9.63	18.60	69.60	2.17
	18.66	4.86	75.34	1.14	9.67	13.94	74.89	1.50

In the upstream, the gastrosomatic index ranged 1.004 to 4.366 in summer and winter season respectively. The value of gastrosomatic index was recorded highest 3.937, 3.708 and 4.366 in winter season and 3.111, 1.434 and 1.454 in summer season. The value was observed lowest 1.004, 1.443 and 1.178 in rainy season (Table 44 and Figure 26a).

Table 44: Seasonal Data of Length, Weight, Alimentary Canal, Relative Length of Gut (RLG) and Gastrosomatic Index (GSI) of Mahseer, *Tor tor* in Upstream of Ganga River

Seasons	*Total Length of Fish (mm)*	*Fish Weight (gms)*	*Length of Alimentary Canal (mm)*	*Weight of Alimentary Canal (gms)*	*Length of Intestine (mm)*	*RLG*	*GSI*
Summer	320	315	730	9.80	112	2.281	3.111
	436	690	1110	9.90	136	2.545	1.434
	340	330	805	4.80	109	2.367	1.454
Rainy	410	478	998	4.80	128	2.434	1.004
	420	658	1030	9.50	119	2.452	1.443
	320	280	728	3.30	99	2.275	1.178
Winter	334	320	787	12.60	107	2.356	3.937
	335	302	795	11.20	102	2.373	3.708
	315	300	678	13.10	95	2.152	4.366

In downstream, the gastrosomatic index was calculated highest 2.815 to 4.151 in winter season whereas it was lowest 0.985 to 1.276 in rainy season of the year. During summer season, the value of gastrosomatic index was recorded 2.327, 1.294 and 1.185 (Table 45).

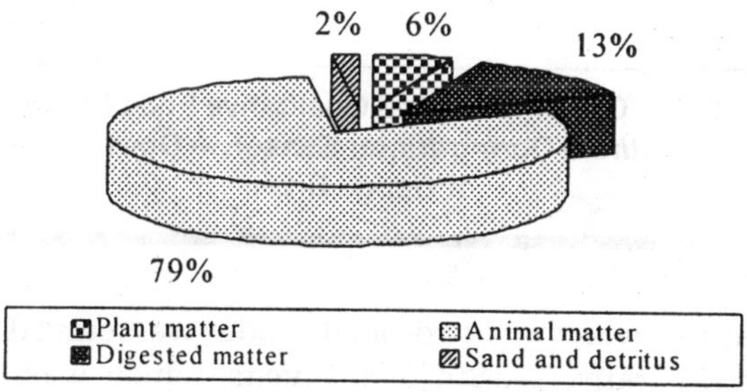

Figure 23(a): Quantitative and Qualitative Food Percentage of *Tor putitora* During Summer Season in Upstream of Ganga River

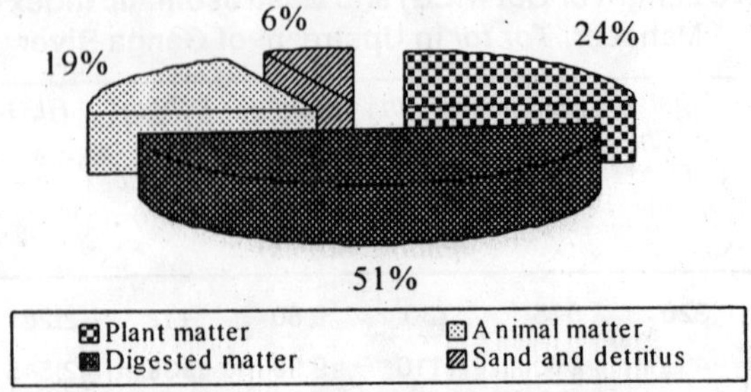

Figure 23(b): Quantitative and Qualitative Food Percentage of *Tor putitora* During Rainy Season in Upstream of Ganga River

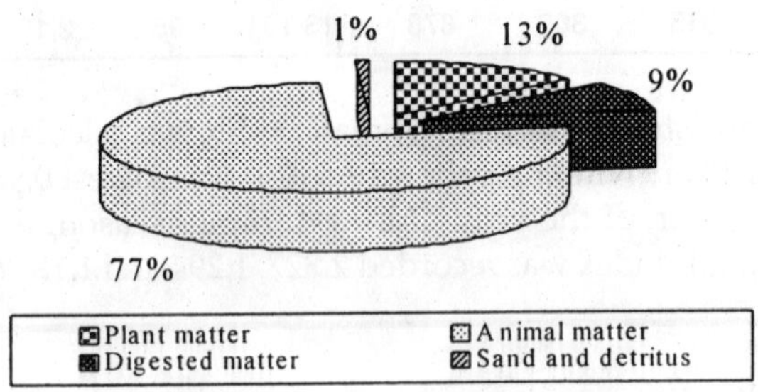

Figure 23(c): Quantitative and Qualitative Food Percentage of *Tor putitora* During Winter Season in Downstream of Ganga River

In Saptsarovar, the gastrosomatic index was found 0.952, 1.550 and 1.263 in rainy season of the year, which was lowest in comparison to summer and winter season. GSI range was found 1.400 to 2.825 in summer season whereas it was recorded highest 4.006, 3.569 and 4.226 in winter season of the year (Table 46 and Figure 26a).

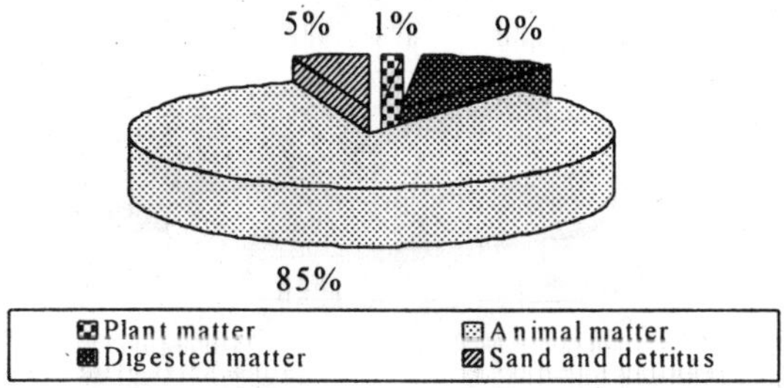

Figure 24(a): Quantitative and Qualitative Food Percentage of *Tor putitora* During Summer Season in Downstream of Ganga River

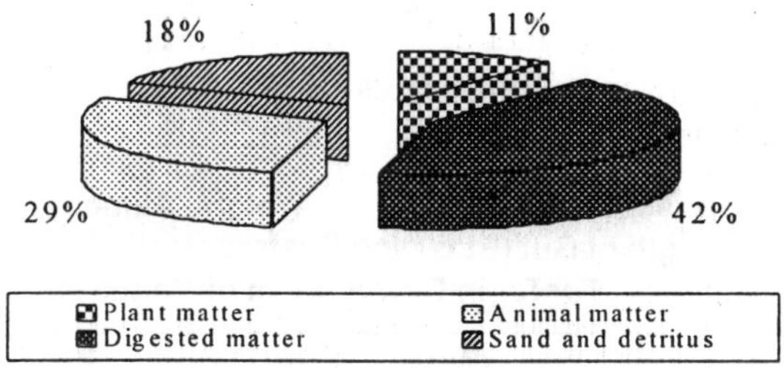

Figure 24(b): Quantitative and Qualitative Food Percentage of *Tor putitora* During Rainy Season in Downstream of Ganga River

Relative Length of Gut (RLG)

The length of intestine varies from fish to fish but in a particular species, it has a definite relation with length of fish, which helps to determine the feeding habits of the fish species. The gastrosomatic index is a relation between weight of alimentary canal to fish weight, which helps in determining the feeding condition in different months and season. But the relative length of gut differs in different stages of

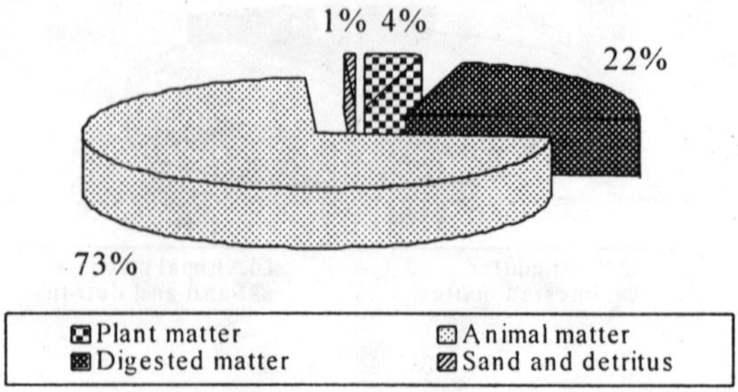

Figure 24(c): Quantitative and Qualitative Food Percentage of *Tor putitora* During Winter Season in Downstream of Ganga River

life history as low in fry, fingerlings and juveniles and high in adults with no seasonal fluctuations.

Table 45: Seasonal Data of Length, Weight, Alimentary Canal, Relative Length of Gut (RLG) and Gastrosomatic Index (GSI) of Mahseer, *Tor tor* in Downstream of Ganga River

Seasons	*Total Length of Fish (mm)*	*Fish Weight (gms)*	*Length of Alimentary Canal (mm)*	*Weight of Alimentary Canal (gms)*	*Length of Intestine (mm)*	*RLG*	*GSI*
Summer	420	610	790	14.20	117	1.880	2.327
	395	680	700	8.80	102	1.772	1.294
	385	565	650	6.70	99	1.688	1.185
Rainy	360	355	602	3.50	97	1.672	0.985
	435	655	835	7.10	109	1.919	1.083
	440	650	850	8.30	112	1.931	1.276
Winter	350	330	580	13.70	98	1.657	4.151
	420	650	805	18.30	117	1.916	2.815
	350	360	582	12.50	105	1.662	3.472

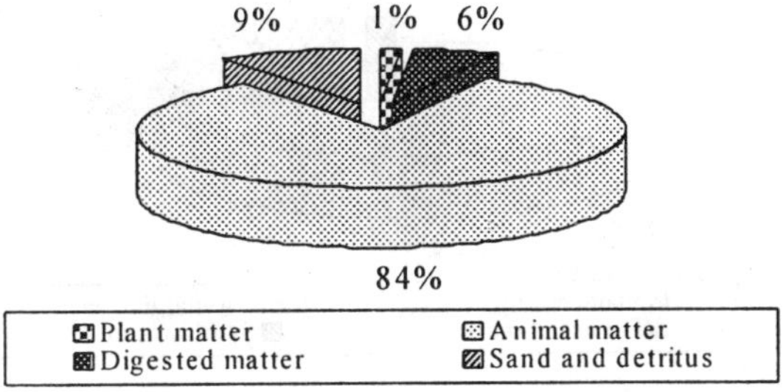

Figure 25(a): Quantitative and Qualitative Food Percentage of *Tor putitora* During Summer Season at Saptsarovar of Ganga River

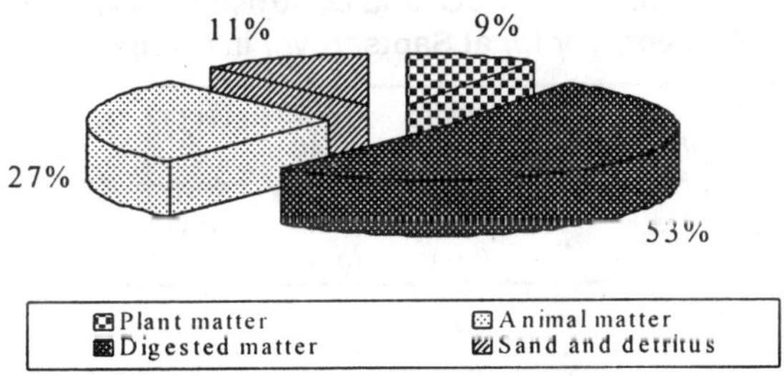

Figure 25(b): Quantitative and Qualitative Food Percentage of *Tor putitora* During Rainy Season at Saptsarovar of Ganga River

The relative gut length of *Tor tor* in upstream of barrage varied from 2.152 to 2.545 in total length range of 315 to 436 mm of fish (Table 44 and Figure 26a). The RLG values 2.281 to 2.545 in total length from 320 to 436 mm in summer followed by 2.275 to 2.452 in total length from 320 to 420 mm in rainy and 2.152 to 2.373 in total length from 315 to 335 mm of *Tor tor* in winter season.

In downstream of barrage, the maximum relative length of gut was calculated 1.931 in total length of 440 mm in rainy season and

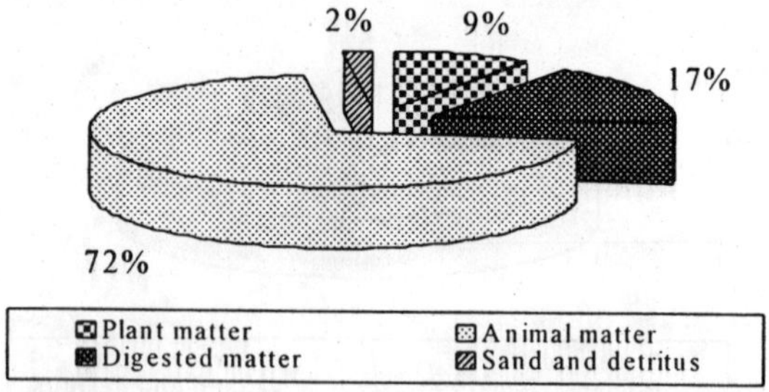

Figure 25(c): Quantitative and Qualitative Food Percentage of *Tor putitora* During Winter Season at Saptsarovar of Ganga River

Table 46: Seasonal Data of Length, Weight, Alimentary Canal, Relative Length of Gut (RLG) and Gastrosomatic Index (GSI) of Mahseer, *Tor tor* at Saptsarovar in Ganga River

Seasons	*Total Length of Fish (mm)*	*Fish Weight (gms)*	*Length of Alimentary Canal (mm)*	*Weight of Alimentary Canal (gms)*	*Length of Intestine (mm)*	*RLG*	*GSI*
Summer	320	315	575	8.90	98	1.796	2.825
	360	425	698	7.20	102	1.938	1.694
	350	400	672	5.60	94	1.920	1.400
Rainy	405	672	805	6.40	107	1.987	0.952
	340	400	622	6.20	98	1.829	1.550
	360	380	695	4.80	102	1.930	1.263
Winter	335	332	630	13.30	105	1.880	4.006
	360	395	700	14.10	110	1.944	3.569
	310	265	540	11.20	98	1.741	4.226.

minimum value was found 1.657 in total length of 350 mm fish in winter season. The RLG values varied from 1.668 to 1.880 in summer months, while 1.672 to 1.931 observed in rainy season and 1.657 to 1.916 found in winter season (Table 45).

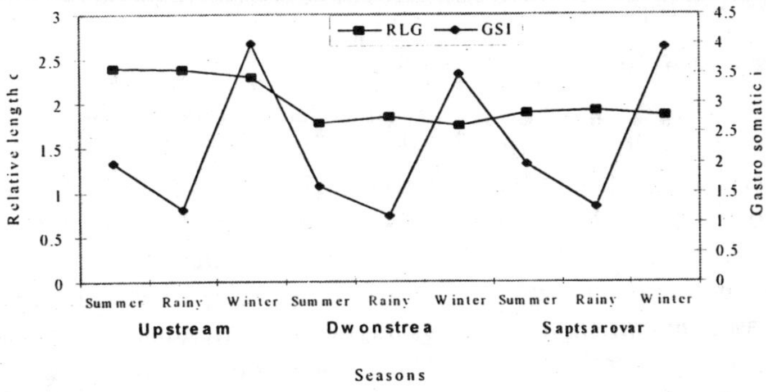

Figure 26(a): Seasonal Variation in Relative Length of Gut and Gastrosomatic Index of *Tor tor* in Upstream, Downstrem and at Saptsarovar in Ganga River

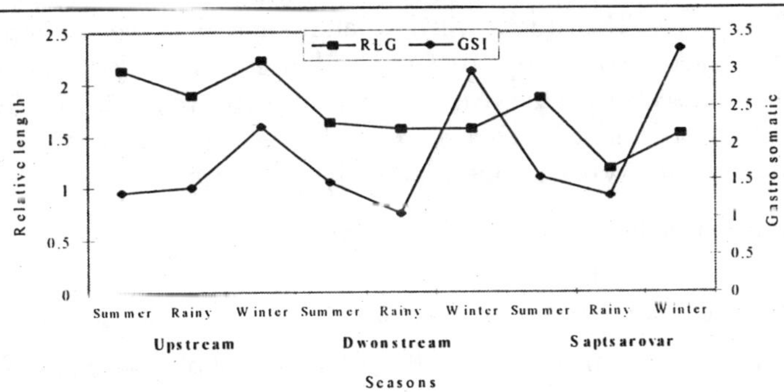

Figure 26(b): Seasonal Variation in Relative Length of Gut and Gastrosomatic Index of *Tor putitora* in Upstream, Downstrem and at Saptsarovar in Ganga River

In Saptsarovar, RLG values varied from 1.741 to 1.987 in tota length from 310 to 405 mm of fish. The maximum value of RLG was observed in rainy season and minimum in winter season of the year (Table 46 and Figure 26a). In the summer season, RLG values were recorded 1.796, 1.938 and 1.920 in total length from 320, 360 and 350 mm of *Tor tor*.

Tor putitora

Gross Morphology of Alimentary Canal

The buccopharynx is also comprised of same organs as in *Tor tor* with similar structure *viz.*, buccal cavity, pharynx, homodont pharyngeal teeth and callous pad (Plate 14). The buccopharynx leads into the muscular oesophagus. The short muscular oesophagus opens through a constriction into a sac like intestinal bulb which become narrowed towards posteroir end. A stomach is absent in *Tor tor* and *Tor putitora* also. After a slightly constriction the bulb leads into a long narrow intestine. The intestine was followed by a short rectum and the anus. The gall bladder opens through bile duct in the middle of the intestinal bulb (Plate15).

Seasonal Qualitative Fluctuations in Food Items

The food items were identified which are given in (Tables 38–40), showing seasonal fluctuations and variations of the food items of *Tor putitora*. The aquatic weeds, diatoms and algae were found in all the season. The group crustacea and insecta were also obtained in all three seasons of the year and taken as more preferred food. The occurrence of minor food items, such as rotifers, protozoans, annelids and fish parts were also observed in the gut of *Tor putitora*.

Quantitative Fluctuations in Food Items

The quantitative food percentage for *Tor putitora* in upstream, downstream of barrage and Saptsarovar are given as follows:

In the upstream of barrage, the percentage of plant matter in gut ranged 4.23 per cent to 7.37 per cent in summer. Highest percentage of the plant matter was recorded 27.52, 30.19 and 14.33 per cent in rainy season whereas, 12.25, 15.21 and 12.45 per cent plant matter was recorded during winter season in gut of *Tor putitora*. Animal matter was found highest 63.69, 55.25 and 32.24 per cent in rainy season while 16.62, 9.82 and 12.21 per cent was observed in summer and lowest percentage was recorded 9.67, 10.32 and 7.43 per cent in winter season. Highest percentage of digested matter ranged 74.51 per cent to 84.03 per cent was observed in summer season and lowest ranged 3.65 per cent to 45.36 per cent recorded in rainy season. Whereas, 77.27, 73.38 and 79.33 per cent of digested matter was found in winter season (Table 41 and Figures 23a–c). Sand and detritus were found in all three season of the year, the percentage was observed lowest 0.79 per cent in winter and highest 8.07 per cent in rainy season in upstream of barrage of Ganga river.

In the downstream, high percentage of plant matter 15.34, 7.97 and 8.75 per cent was found in rainy season and 3.15, 5.89 and 2.21 per cent was found in winter season. The lowest percentage 1.11, 0.78 and 2.10 per cent was recorded in winter season. Animal matter was observed highest ranged 30.84 per cent to 61.11 per cent in rainy season followed by 16.27, 24.99 and 24.75 per cent in winter season. The lowest percentage 6.50 per cent to 13.76 per cent of animal matter was found in summer season. The percentage of digested matter was highest 78.98 per cent to 88.93 per cent in summer followed by 68.93 per cent to 80.05 per cent in winter season. The lowest percentage 3.94, 40.25 and 41.65 per cent of digested matter were found in gut during rainy season. Sand and detritus were recorded maximum 19.61 per cent in rainy season and minimum 0.53 per cent in winter season (Table 42 and Figures 24a–c).

In Saptsarover, the percentage of plant matter was 10.81, 6.93 and 8.85 per cent were observed in rainy season. 8.26, 9.63 and 9.67 per cent of plant matter was found in winter season and lowest percentage 1.35, 1.98, and 1.00 per cent were recorded in summer season. High percentage of animal matter 58.94, 55.48 and 43.45 per cent were observed in rainy season and lowest percentage were 9.11, 2.29 and 6.23 per cent found in summer season. During the winter season, animal matter ranged 13.94 per cent to 19.78 per cent was observed in gut. Digested matter percentage was highest 80.57, 84.11 and 85.20 found in summer and lowest 19.72, 24.26 and 37.07 per cent ages were recorded in rainy season. Digested matter percentages in winter season were observed 69.77, 69.60 and 74.89 per cent. Sand and detritus percentage were found highest in all three season in gut of *Tor putitora* due to its coulmn feeding nature (Table 43 and Figure 25a–c).

Gastrosomatic Index (GSI)

In the upstream, the gastrosomatic index ranged from 0.585 in summer to 2.418 in winter season. After rainy season, the gastrosomatic index was observed highest. In rainy season the gastrosomatic index were recorded 1.157, 1.254 and 1.825 (Table 47 and Figure 26b).

Table 47: Seasonal Data of Length, Weight, Alimentary Canal, Relative Length of Gut (RLG) and Gastrosomatic Index (GSI) of Mahseer, *Tor putitora* in Upstream of Ganga River

Seasons	*Total Length of Fish (mm)*	*Fish Weight (gms)*	*Length of Alimentary Canal (mm)*	*Weight of Alimentary Canal (gms)*	*Length of Intestine (mm)*	*RLG*	*GSI*
Summer	310	700	600	15.10	89	1.935	2.157
	390	540	830	6.80	90	2.128	1.259
	425	700	950	4.10	52	2.325	0.585
Rainy	425	700	945	8.10	74	2.223	1.157
	252	555	340	3.20	54	1.349	1.254
	368	400	770	7.30	61	2.0292	1.825
Winter	390	550	840	13.30	89	2.153	2.418
	310	480	615	10.50	78	1.983	2.187
	480	910	1220	19.10	91	2.541	2.098

In downstream, the gastrosomatic index was observed lowest 0.780, 1.40 and 1.00 in rainy season whereas; it was highest 3.219, 3.500 and 2.187 in winter season, just after spawning period. Before spawning or pre-spawning period GSI was decreased 1.914 to 1.155 in summer season (Table 48).

In Saptsarovar, the gastrosomatic index was calculated highest 3.090, 3.793 and 2.918 in winter season. The GSI was observed 2.212, 1.114 and 1.297 in summer and 1.400, 1.323 and 1.164 recorded in rainy season. The GSI was observed minimum 1.114 in pre-spawning period and maximum 3.793 recorded in post spawning period of fish (Table 49 and Figure 26b).

Relative Length of Gut (RLG)

The relative length of gut calculated and varied from 0.891 to 1.877 in total length range 310 to 480 mm of fish (Table 47 and Figure 26b). The general picture of the relative length of gut shows that the RLG values of *Tor putitora* in upstream of barrage varied from 1.349 to 2.541 in total length from 252 mm to 480 mm of fish. The RLG values in summer season were recorded 1.935 to 2.325 followed by 1,349 to 2.223 in rainy season and 1.983 to 2.541 were

Table 48: Seasonal Data of Length, Weight, Alimentary Canal, Relative Length of Gut (RLG) and Gastrosomatic Index (GSI) of Mahseer, *Tor putitora* in Downstream of Ganga River

Seasons	*Total Length of Fish (mm)*	*Fish Weight (gms)*	*Length of Alimentary Canal (mm)*	*Weight of Alimentary Canal (gms)*	*Length of Intestine (mm)*	*RLG*	*GSI*
Summer	465	700	960	13.40	79	2.064	1.914
	300	350	405	4.80	58	1.350	1.371
	356	450	525	5.20	72	1.474	1.155
Rainy	375	500	590	3.90	68	1.573	0.780
	310	250	425	3.50	49	1.370	1.400
	410	600	715	6.05	85	1.743	1.008
Winter	380	410	630	13.20	77	1.657	3.219
	234	100	280	3.50	38	1.196	3.500
	410	480	755	10.50	87	1.841	2.187

Table 49: Seasonal Data of Length, Weight, Alimentary Canal, Relative Length of Gut (RLG) and Gastrosomatic Index (GSI) of Mahseer, *Tor putitora* Saptsarovar in Ganga River

Seasons	*Total Length of Fish (mm)*	*Fish Weight (gms)*	*Length of Alimentary Canal (mm)*	*Weight of Alimentary Canal (gms)*	*Length of Intestine (mm)*	*RLG*	*GSI*
Summer	300	235	498	5.20	68	1.660	2.212
	390	610	798	6.80	79	2.046	1.114
	365	370	690	4.80	68	1.890	1.297
Rainy	320	300	535	4.20	50	1.671	1.400
	350	340	660	4.50	81	1.885	1.323
	325	262	548	3.05	50	1.686	1.164
Winter	300	220	500	6.80	58	1.666	3.090
	264	145	320	5.50	49	1.212	3.793
	320	305	538	8.90	62	1.681	2.918

recorded winter season of the year. The value of RLG was highest in upstream because more plant matter were found than animal matter.

The RLG value in downstream of barrage varied from 1.196 to 2.064 in total length from 234 mm to 465 mm of *Tor putitora*. The RLG value was observed lowest in downstream fishes because more animal matter were found in downstream of barrage than plant matter. The values of RLG were calculated 1.350 to 2.064 for 300 mm to 465 mm fishes while, 1.370 to I.743 were recorded for 310 mm and 410 mm fishes in summer and rainy season respectively. RLG values were recorded 1.196 to 1.841 for 234 mm to 410 mm of *Tor putitora* in winter season (Table 48).

In Saptsarovar, RLG values were observed 1.660 to 2.046 for total length from 300 to 390 mm of fishes for summer, 1.671 to 1.885 in total length 320 to 350 mm of fishes for rainy season and 1.212 to 1.681 RLG were calculated in total length from 264 to 320 mm of *Tor putitora* in winter months (Table 49 and Figure 26b).

Sexual Dimorphism and Sex Ratio

Tor tor

Colour

The body colour of male and female does not appear to distinguish the sexes, although in some males the colour is brighter than the females. Some males also have small black spots on the lateral sides of mouth, which are absent in females.

Size

The males were found smaller in size and less in weight than female specimens of same age group. The head length and body depth relation (head length/ body depth) in male is higher than females of the same age during the breeding seasons.

Fins

During breeding season, the pectoral fins in males are comparatively longer and rough in nature and reaches up to the sixth scales of the lateral line. But in females the pectoral fin reach, below fifth scales only of lateral line and are smooth in nature. The anal fin of male is thinner, transparent and larger in size with light orange ting at the tip, while it is thicker, opaque and smaller in size in female of about same age.

Tubercles

Many small tubercles are present on operculum bone of both sides, which are also present on the dorsal side in the head region in males.

Tor putitora

Sexual characters have also been recorded on *Tor putitora* as follows:

Colour

In the fish, the mature male has brighter colouration of body than the female, while this differentiation may not by sharply marked, it is helpful when taken with other characters during breeding seasons. The anal fin has orange tinge at the tip in both sexes but it has brighter orange colour in the female than in the male. This character is only applicable after examining a large number of specimens.

Size

In this view, the male is smaller in size than the female of the same age. In the ripe females, the body depth is about equal to head length while in ripe male the head length is comparatively larger than the depth of body. The character usually clinches the differentiation of males from the females.

Fins

The anal fin is opaque and smaller in size than the pectoral in female whereas it is equal in length to pectoral fin and is more transparent in male specimens. The pectoral fin is rough in males and more or less smooth in females. The length of pectoral fin, however is a fluctuating character and cannot by used for sex differentiation as can be done for *Tor tor*.

Tubercles

Contrary to *Tor tor*, no tubercles were observed in male specimens of *Tor putitora* in the present study.

Other Characters

A prominent outgoing body streak arises during breeding seasons in males, which starts anterior to the tip of maxillary bone (upper jaw) and runs underneath the eye ball up to opercular bone on both sides of the head. This streak is not present in females of *Tor putitora* and is also absent in the both sexes of *Tor tor* (Table 50 and Plate 16).

Plate 16: Mahseer, *Tor putitora* (Male and Female Specimen)

The sex ratio have observed in *Tor tor* indicate that more males were present in Ganga river than females, as observed 1 : 1.345 ratio for upstreanı of barrage, 1 : 1.642 in downstream and 1 : 1 in Saptsarovar. The highest numbers of males were found in rainy season (75 per cent) and the high value (70 per cent)was also recorded in summer season, when few males were chasing to a single female for breeding. There were total 74 male and 55 females were observed in upstream, and the sex ratio was highest 1 : 3 during the rainy season of the study period (Table 51 and Figure 27a).

Table 50: Differentiation of Secondary Sexual Dimorphism of *Tor tor* and *Tor putitora* in Breeding Seasons

Characters	*Tor tor*		*Tor putitora*	
	Male	*Female*	*Male*	*Female*
Body size	Shorter Similar	Longer Similar	Alike	Alike
Snout tubercles	Many during Breeding Seasons	Nil	Few	Few
Lips	More protruded	Less protruded	Thicker protruded	Less protruded
Dorsal spine	Similar	Similar	Thicker & Longer	Thicker & Short
Pectoral fin	Similar in size, colour	Similar in size, colour	Bright orange rims	Slight pink
Anal fins	Greyish	Greyish	Reddish	Pinkish

In downstream, sex ratio was observed 1 : 2 in summer season of the year when breeding is just started. The percentage of male

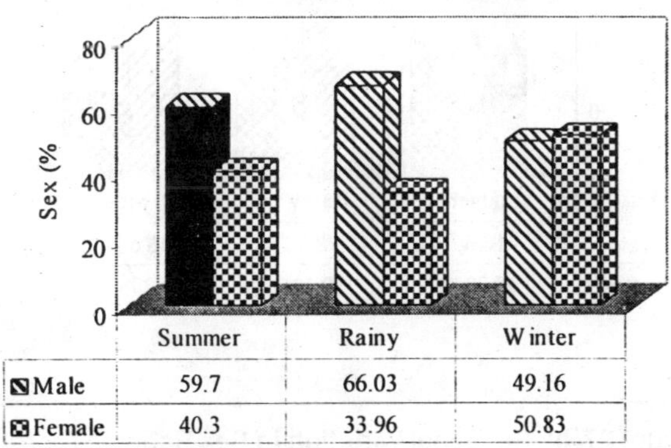

Figure 27(a): Seasonal Variation in Sex Ratio of *Tor tor* in Upstream of Ganga River

Table 51: Seasonal Data of Sex Ratio in Mahseer, *Tor tor* in Upstream of Ganga River

Seasons	*Total. No of Fish*	*Male*		*Female*		*Sex Ratio*		*Expected Number*		*Chi Sqaure*
		No.	*%age*	*No.*	*%age*	*Male*	*Female*	*Male*	*Female*	*(χ^2)*
Summer	18	8	44.40	10	55.60	0.80	1.00	9.00	9.00	0.22
	17	11	64.70	6	35.30	1.83	1.00	8.50	8.50	1.47
	10	7	70.00	3	30.00	2.33	1.00	5.00	5.00	1.60
Rainy	12	9	75.00	3	25.00	3.00	1.00	6.00	6.00	3.00
	14	9	64.30	5	35.70	1.80	1.00	7.00	7.00	1.14
	17	10	58.80	7	41.20	1.43	1.00	8.50	8.50	0.53
Winter	16	7	43.70	9	56.30	0.78	1.00	8.00	8.00	0.25
	12	6	50.00	6	50.00	1.00	1.00	6.00	6.00	0.00
	13	7	53.80	6	46.20	1.17	1.00	6.50	6.50	0.07
Total	**129**	**74**	**58.30**	**55**	**41.70**	**14.14**	**9.00**	**64.50**	**64.50**	**8.28***

* Significant at 0.05

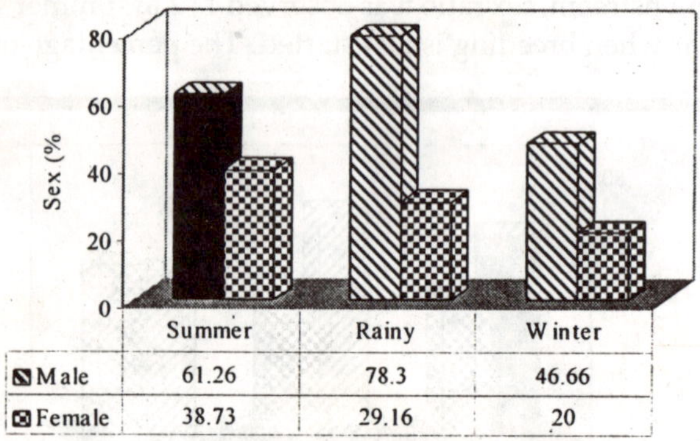

Figure 27(b): Seasonal Variation in Sex Ratio of *Tor tor* in Downstream of Ganga River

was recorded 66.7 per cent during this season. Sex ratio was recorded 0 : 0 in few months of the winter and last month of the rainy season, when breeding was completed (Table 52 and Figure 27b).

Table 52: Seasonal Data of Sex Ratio in Mahseer, *Tor tor* in Downstream of Ganga River

Seasons	Total. No of Fish	Male		Female		Sex Ratio		Expected Number		Chi Sqaure
		No.	%age	No.	%age	Male	Female	Male	Female	(χ^2)
Summer	5	3	60.00	2	40.00	1.50	1.00	2.50	2.50	0.20
	3	2	66.70	1	33.30	2.00	1.00	1.50	1.50	0.33
	7	4	57.10	3	42.90	1.33	1.00	3.50	3.50	0.14
Rainy	8	5	62.30	3	37.50	1.67	1.00	4.00	4.00	0.50
	4	2	50.00	2	50.00	1.00	1.00	2.00	2.00	0.00
	3	3	100.00	0	0.00	0.00	0.00	1.50	0.00	1.50
Winter	5	2	40.00	3	60.00	0.67	1.00	2.50	2.50	0.20
	2	2	100.00	0	0.00	0.00	0.00	1.00	0.00	1.00
	0	0	0.00	0	0.00	0.00	0.00	0.00	0.00	0.00
Total	**37**	**23**	**59.58**	**14**	**29.30**	**8.17**	**6.00**	**18.50**	**16.00**	**3.87**

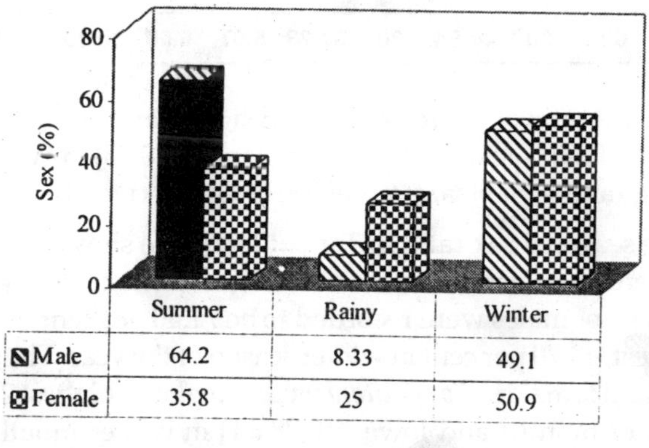

Figure 27(c): Seasonal Variation in Sex Ratio of *Tor tor* at Saptsarovar of Ganga River

In Saptsarovar, the highest percentages of males were recorded to be 100 per cent in summer season. The Sex ratio was recorded 0 : 0 in summer months and 1 : 1.25 in winter season of the year.

Minimum percentage of male was found to be 0-25 per cent in rainy season and the percentage shown increasing trend upto summer months (Table 53 and Figure 27c).

Table 53: Seasonal Data of Sex Ratio in Mahseer, *Tor tor* at Saptsarovar in Ganga River

Seasons	*Total. No of Fish*	*Male*		*Female*		*Sex Ratio*		*Expected Number*		*Chi Sqaure*
		No.	*%age*	*No.*	*%age*	*Male*	*Female*	*Male*	*Female*	(χ^2)
Summer	19	10	52.60	9	47.40	1.11	1.00	9.50	9.50	0.05
	5	2	40.00	3	60.00	0.67	1.00	2.50	2.50	0.20
	3	3	100.00	0	0.00	0.00	0.00	1.50	0.00	1.50
Rainy	0	0	0.00	0	0.00	0.00	0.00	0.00	0.00	0.00
	0	0	0.00	0	0.00	0.00	0.00	0.00	0.00	0.00
	4	1	25.00	3	75.00	0.33	1.00	2.00	2.00	1.00
Winter	8	4	50.00	4	50.00	1.00	1.00	4.00	4.00	0.00
	9	5	55.60	4	44.40	1.25	1.00	4.50	4.50	0.11
	12	5	41.70	7	58.30	0.71	1.00	6.00	6.00	0.33
Total	**60**	**30**	**40.54**	**30**	**37.23**	**5.07**	**6.00**	**30.00**	**28.50**	**3.19**

The sex ratio of *Tor tor* was found significant ($p < 0.05$) only in upstream of barrage while non-significant ($p > 0.05$) in downstream of barrage and at Saptsarovar region of Ganga river.

The seasonal sex ratio in *Tor putitora* were showed that more males were present in upstream of barrage than females. The highest percentage of males were recorded to be 71.40 per cent in summer and lowest, 43.70 per cent in winter season of the year. The sex ratio of male and female in *Tor putitora* were calculated highest 2.50 : 1.00 in summer months and lowest (0.78 : 1) in winter months of the study period (Table 54 and Figure 28a).

In the downstream, the average sex ratio was recorded 1.00 : 1.435 between females and males, which indicate that more population of males in downstream of barrage. The highest percentage (100 per cent) of male was observed in winter season of the year. The sex ratio of female and male were calculated 0 : 0 in winter months when more males were recorded in the river (Table 55 and Figure 28b).

In Saptsarovar, the average sex ratio between female and males were 1 : 1.083 in both the years. The ratio between females and males

Table 54: Seasonal Data of Sex Ratio in Mahseer, *Tor putitora* in Upstream of Ganga River

Seasons	*Total. No of Fish*	*Male*		*Female*		*Sex Ratio*		*Expected Number*		*Chi Sqaure* (χ^2)
		No.	*%age*	*No.*	*%age*	*Male*	*Female*	*Male*	*Female*	
Summer	16	9	56.30	7	43.70	1.29	1.00	8.00	8.00	0.25
	17	10	58.80	7	41.20	1.43	1.00	8.50	8.50	0.53
	14	10	71.40	4	28.60	2.50	1.00	7.00	7.00	2.57
Rainy	10	7	70.00	3	30.00	2.33	1.00	5.00	5.00	1.60
	20	13	65.00	7	35.00	1.86	1.00	10.00	10.00	1.80
	16	10	62.50	6	37.50	1.67	1.00	8.00	8.00	1.00
Winter	25	12	48.00	13	52.00	0.92	1.00	12.50	12.50	0.04
	20	9	45.00	11	55.00	0.82	1.00	10.00	10.00	0.20
	16	7	43.70	9	56.30	0.78	1.00	8.00	8.00	0.25
Total	**154**	**87**	**57.85**	**67**	**42.14**	**13.60**	**9.00**	**77.00**	**77.00**	**8.24***

* Significant at 0.05

Table 55: Seasonal Data of Sex Ratio in Mahseer, *Tor putitora* in Downstream of Ganga River

Seasons	*Total. No of Fish*	*Male*		*Female*		*Sex Ratio*		*Expected Number*		*Chi Sqaure* (χ^2)
		No.	*%age*	*No.*	*%age*	*Male*	*Female*	*Male*	*Female*	
Summer	11	5	45.40	6	54.60	0.83	1.00	5.50	5.50	0.09
	13	8	61.50	5	38.50	1.60	1.00	6.50	6.50	0.69
	6	4	66.70	2	33.30	2.00	1.00	3.00	3.00	0.67
Rainy	16	10	62.50	6	37.50	1.67	1.00	8.00	8.00	1.00
	15	11	73.30	4	26.50	2.75	1.00	7.50	7.50	3.27
	8	5	62.50	3	37.50	1.67	1.00	4.00	4.00	0.50
Winter	16	7	43.70	9	56.30	0.78	1.00	8.00	8.00	0.25
	9	5	55.60	4	44.40	1.25	1.00	4.50	4.50	0.11
	1	1	100.00	0	0.00	0.00	0.00	0.50	0.00	0.50
Total	**95**	**56**	**63.46**	**39**	**36.51**	**12.55**	**8.00**	**47.50**	**47.00**	**7.08**

were found. highest 1 : 3 in summer season of the year when breeding was started, the highest percentage (75.0 per cent) of males were also recorded in similar season. The percentage of males were recorded lowest (33.3 per cent) in rainy season of the year (Table 56 and Figure 28c).

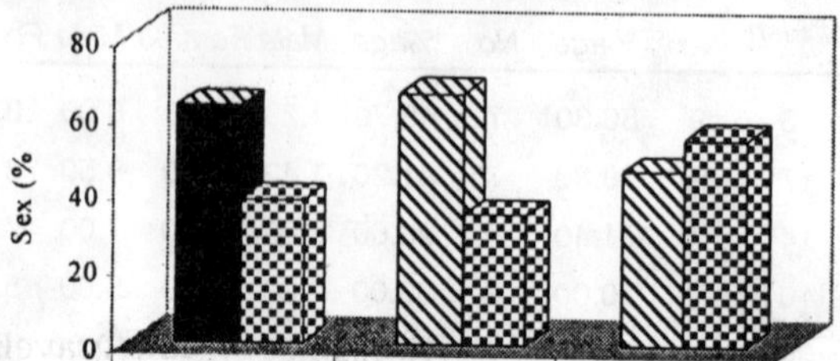

	Summer	Rainy	Winter
Male	62.16	65.83	45.56
Female	37.83	34.16	54.43

Figure 28(a): Seasonal Variation in Sex Ratio of *Tor putitora* in Upstream of Ganga River

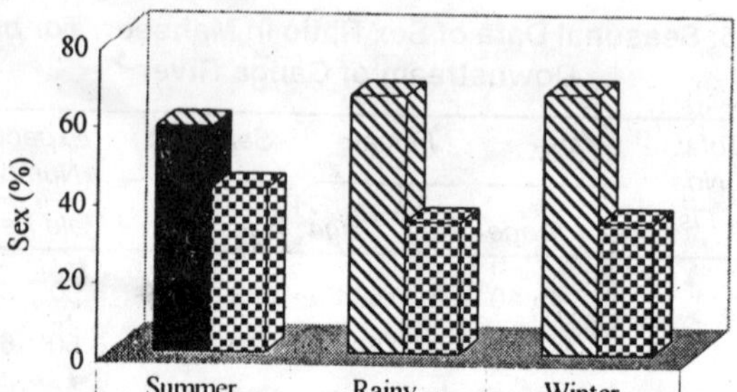

	Summer	Rainy	Winter
Male	57.86	66.1	66.43
Female	42.13	33.83	33.56

Figure 28(b): Seasonal Variation in Sex Ratio of *Tor putitora* in Downstream of Ganga River

Table 56: Seasonal Data of Sex Ratio in Mahseer, *Tor putitora* at Saptsarovar in Ganga River

Seasons of Fish	*Total. No*	*Male*		*Female*		*Sex Ratio*		*Expected Number*		*Chi Sqaure* (χ^2)
		No.	*%age*	*No.*	*%age*	*Male*	*Female*	*Male*	*Female*	
Summer	9	5	55.60	4	44.40	1.25	1.00	4.50	4.50	0.11
	4	3	75.00	1	25.00	3.00	1.00	2.00	2.00	1.00
	6	3	50.00	3	50.00	1.00	1.00	3.00	3.00	0.00
Rainy	8	5	62.50	3	37.50	1.67	1.00	4.00	4.00	0.50
	5	2	40.00	3	60.00	0.67	1.00	2.50	2.50	0.20
	3	1	33.30	2	66.70	0.50	1.00	1.50	1.50	0.33
Winter	12	6	50.00	6	50.00	1.00	1.00	6.00	6.00	0.00
	15	8	53.30	7	46.70	1.14	1.00	7.50	7.50	0.06
	13	6	46.20	7	53.80	0.85	1.00	6.50	6.50	0.07
Total	**75**	**39**	**51.76**	**36**	**48.23**	**11.08**	**9.00**	**37.50**	**37.50**	**2.27**

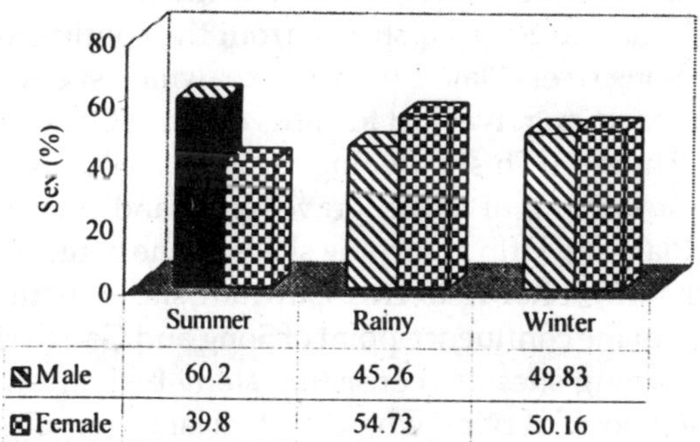

Figure 28(c): Seasonal Variation in Sex Ratio of *Tor putitora* at Saptsarovar of Ganga River

Thus, sex ratio in between female and male Tor putitora was only significant ($p < 0.05$) in upstream of barrage and the ratio was non-significant ($p > 0.05$) in downstream of barrage and Saptsarovar.

Location of Natural Spawning Grounds

The natural spawning sites were observed as the collection of mature fertilized eggs from bottom of small pebbles, gravels and stones substratum, containing with aquatic macrophytes. Spawning of mahseer sites were surveyed in Ganga river upto two years. The map (Figure 29) shows the location of spawning grounds and spawning sites of mahseer in upstream, downstream and Saptsarovar region of Ganga river and its tributaries like Hanval, Song and Suswa river at Rishikesh region. The spawning grounds of mahseer are shallow banks of the river and pools containing 35 to 50 cm depth with less gradient.

First spawning site have an area about 2.0 to 5.0 meters length and 0.5 to 1.0 meter water depth containing a very small pebbles and gravels (Plate 17). This site is situated just towards downstream of confluence point of Hanval and Ganga river near Phoolchatti Ashram (about 10 km upstream of Rishikesh). Second spawning site is situated on the right bank of Ganga river near Ved Niketan Ashram (downwards of Laxmanjhula) at Rishikesh (Plate 18). Third spawning site is on the bank of Suswa river near Midway Restaurant Raiwala, about 4 to 5 km upstream from the confluence point of Suswa and Song river (Plate 19). Fourth spawning site is selected on the right bank of Song river just towards down of Doiwala bridge at Doiwala, Dehradun. This spawning site is broad but in destructing phase due to mining of stone, gravel and sand for construction purposes (Plate 27). Fifth spawning site is on the bank of Song river near Nepali farm at Raiwala. This spawning site is about 5 to 10 km upwards from the confluence point of Song and Ganga river (Plate 20). All spawning sites are consisting stony bed substratum and shallow deep pools of river, where the fish laid eggs and fry.

The above spawning sites having shallow region and pools of confluence point of Song and Suswa, Song and Ganga and Hanval and Ganga or some time upward of confluence point. All the spawning sites were observed in most prevailing favorable condition for breeding of mahseer. The spawning grounds contain high water temperature, dissolved oxygen and ground feeding materials. The bottom of spawning grounds were pebbly, gravely with sand, which is a favorable fish-spawning substratum. Many fertilized eggs, larvae and fry were collected from these regions by a long handle plankton dip net (Plate 21 and 23). Eggs and larvae were collected mainly

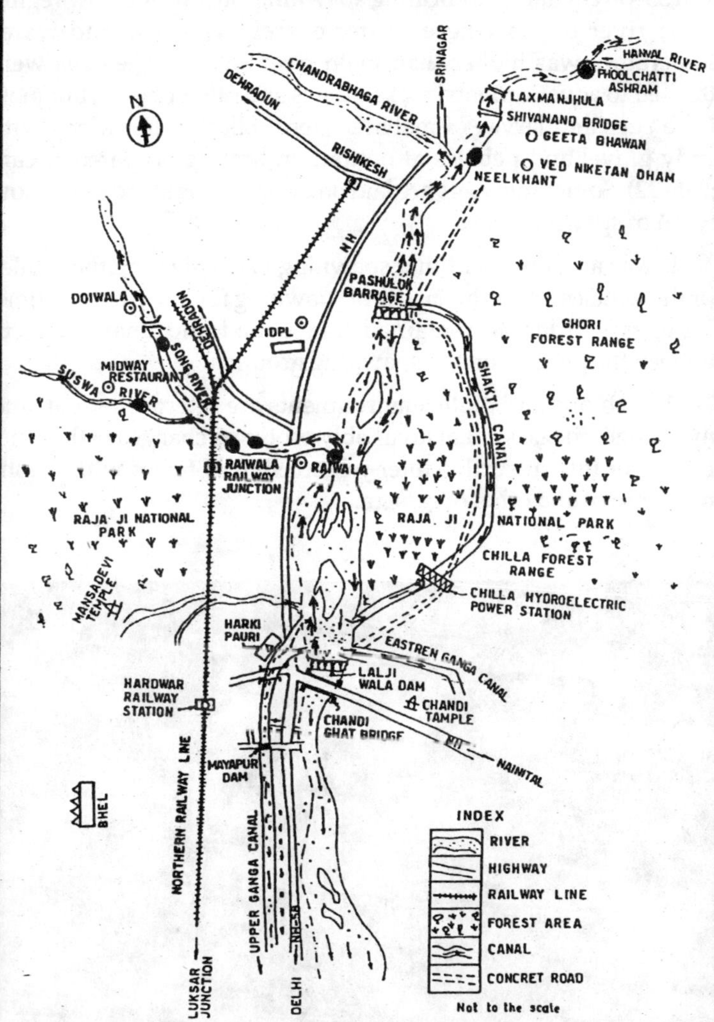

Figure 29: Map of Ganga River and its Tributaries Showing Spawning Sites (●) and Migration Route of Mahseer (→) (*Tor tor* and *Tor putitora*)

from 35–50 cm depth, indicating spawning sites are shallow region of the river banks where, water current was low and water temperature was higher than main river stream. The eggs were attached to aquatic plants, rocks, stones and other objects, but most of the eggs and larvae of mahseer were collected free along with sandy gravel and pebbles at bottom, indicating no parental care (Plate 22). Sometimes eggs are laid on or under the stone in shallow region of spawning sites.

It was also observed that spawning takes place neither males nor the females stayed at the site of spawning for any length of time. The eggs and larvae remained unattached and unattended in between the pebbles on the spawning grounds.

The abiotic and biotic environments are interdependent, and any changes in one system invariably produces changes in the other because of this subjective inter-relationship of fishes with abiotic environment and other organism.

Plate 17: Showing Spawning Site in confluence point of Hanval and Ganga River near Phoolchatti Ashram at Rishikesh

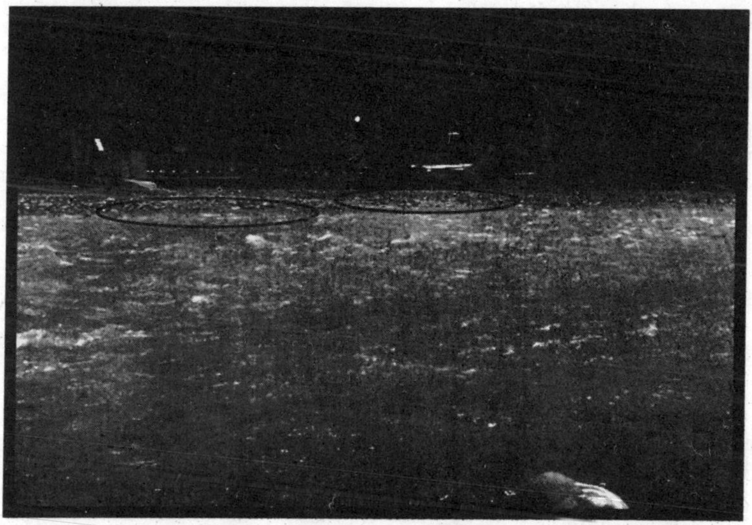

Plate 18: Showing Location of Spawning Site near Ved Niketan Ashramat Rishikesh

Plate 19: Showing Location of Spawning Site of Mahseer in Song River

Plate 20: Showing Location of Spawning Site of Mahseer near the Confluence point of Song and Ganga River at Saptsarovar

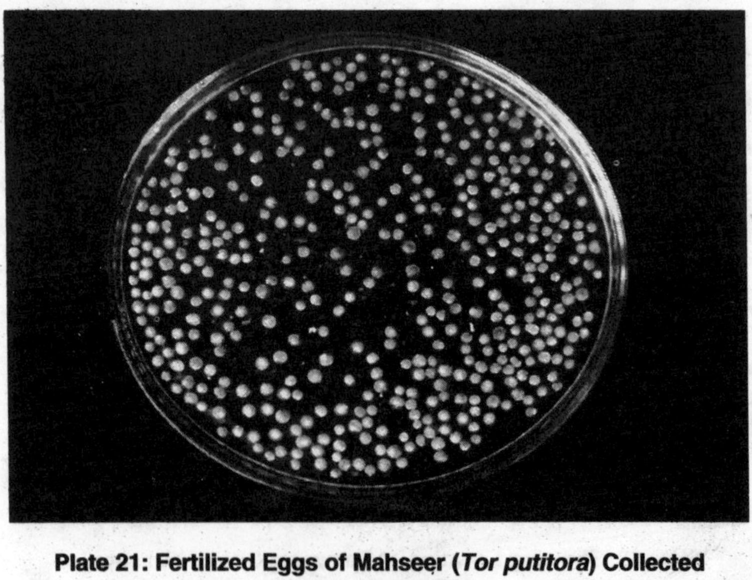

Plate 21: Fertilized Eggs of Mahseer (*Tor putitora*) Collected from Spawning Sites

Plate 22: Fertilized Eggs of Mahseer Attached with Stone and Aquatic Vegetations

Plate 23: Mahseer's Fry Swimming in Ganga River

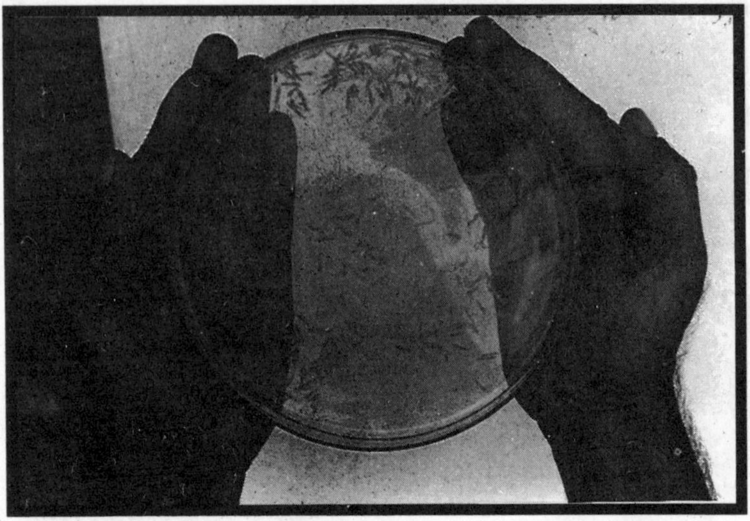

Plate 24: Hatchling of Mahseer hatched in Controlled Developing Condition

Plate 25: Swin Up Fry of Mahseer

Plate 26: Fingerlings of Mahseer Collected from Vyasghat

Plate 27: Showing Extraction Activity of Stone, Pebbles and Gravals from Ganga River, Destroying the Natural Feeding and Breeding Grouds of Mahseer

The most important ecological factor influencing the whole eco-biology of mahseer fishes. During the present study, some important physico-chemical parameters like water temperature, velocity, pH, dissolved oxygen, hardness and turbidity etc. and biological parameters like plankton and macro-benthic population of spawning grounds of mahseer were analysed and given in (Tables 57, 58 and Figures 30 & 31). During spawning seasons (April to September) the water temperature ranged 18.55±0.872 to 26.075±2.407°C, water velocity from 61.50±7.889 to 110.333±8.710 cm/sec and dissolved oxygen from 7.516±0.307 to 9.70±0.187 mg/l were analysed. Lowest values of dissolved oxygen 7.516±0.3 70 mg/l recorded during rainy season (July to September), may be attributed to the washing of inorganic matters due to rains and decreased photosynthetic activity caused by high turbidity.

Hardness was observed minimum 167.5±5.590 mg/l in April to May at site V and calcium was 17.275±0.465 mg/l at site I in similar months. While, chloride concentration was analysed maximum 16.25±l.675 mg/l in July to September at site II, due to rainy season and minimum 11.875±0.257 mg/l in April to May at site III (Table 57 and Figure 30).

The lotic environments are allows low plankton and benthic fauna due to high turbidity and water current. In present study, the planktonic population varied and recorded as zooplankton 23.833±23.819 nos/ml to 245.00±51.599 nos/ml and phytoplankton 267.5±59.476 nos/ml to 1117.5±352.357 nos/ml in the spawning grounds of mahseer (Table 58 and Figures 31a&b). The macro-benthic population were found maximum 29.0±10.954 ind/m^2 in the months of April to May at site V and minimum 6.0±4.753 ind/m^2 in July to September months at site I. Population of the planktons and macrobenthic fauna were minimum in July to September months due to high turbidity and high current by the rains.

Some fertilized eggs were collected from these spawning sites and reared in wooden trays (size 60 × 30 × 10 cm) with sand and pebbles. The diameter of collected eggs were recorded 2.8 to 3.0 mm in April-May and 2.9 to 3.2 mm were recorded in July-September batches. All the collected fertilized eggs were hatched 85.5 to 94.0 hours at 23.4°C and 82.0–90.2 hours at 24.9°C water temperature in these trays under controlled condition. The cumulative survival rates calculated 93.20 per cent and 87.45 per cent under controlled

Table 57: Physico-chemical Characteristics of Spawning Grounds (with ± S.D.) of Mahseer in Ganga River and its Tributaries

Physico-chemical Parameters	*Site–I*		*Site–II*	
	April to May	*July to September*	*April to May*	*July to September*
Temperature (°C)	21.45±0.837 (20.1–22.4)	24.35±0.947 (22.9–25.4)	18.55±0.872 (17.2–19.5)	23.00±2.060 (19.7–25.1)
Turbidity (NTU)	8.30±0.570 (7.5–9.1)	18.35±1.431 (16.4–20.2)	9.05±1.409 (7.6–11.2)	17.78±1.239 (16.1–19.3)
Velocity (cm/sec)	61.50±7.889 (50–71)	103.33±12.355 (90–122)	68.00±9.137 (58–81)	93.00±7.924 (85–103)
pH	8.25±0.228 (7.9–8.5)	8.23±0.391 (7.8–8.8)	7.85±0.426 (7.3–8.5)	7.93±0.455 (7.4–8.5)
Dissolved Oxygen (mg/l)	9.70±0.187 (9.4–9.9)	8.93±0.431 (8.3–9.5)	8.62±0.395 (8.4–9.1)	8.16±0.545 (7.4–8.9)
Hardness (mg/l)	106.25±6.647 (98–115)	115.66+21.087 (89–145)	93.00±6.652 (86–104)	98.5±12.708 (84–117)
Calcium (mg/l)	17.27±0.465 (16.6–17.9)	14.98±0.807 (13.9–16.3)	15.82±0.878 (14.4–16.8)	13.46±l.194 (12.3–15.1)

Physico-chemical Parameters	*Site–III*		*Site–IV*	
	April to May	*July to September*	*April to May*	*July to September*
Temperature (°C)	26.07±2.407 (23.1–29.0)	25.73±0.673 (24.9–26.4)	25.47±2.706 (22.6–29.0)	25.83±0.830 (24.8–26.9)
Turbidity (NTU)	7.45±0.601 (6.8–8.4)	14.01±1.018 (12.4–15.3)	8.12±0.553 (7.4–8.8)	19.13±3.420 (14.8–17.7)
Velocity (cm/sec)	64.00±7.382 (54–74)	86.16±12.859 (71–105)	75.50±87.89 (65–88)	110.33±8.710 (95–121)
pH	7.85±0.303 (7.5–8.2)	7.86±0.454 (7.4–8.4)	7.47±0.145 (7.3–7.7)	7.60±0.349 (7.1–8.0)
Dissolved Oxygen (mg/l)	8.22±0.215 (7.9–8.5)	7.66±0.515 (7.0–8.3)	8.50±0.430 (7.9–9.0)	7.51±0.370 (7.0–8.0)
Hardness (mg/l)	115.75±4.968 (108–121)	101.33±5.64 (94–110)	163.50±3.640 (158–168)	111.83±8.541 (101–124)
Calcium (mg/l)	14.47±0.480 (13.8–15.1)	11.46±0.503 (10.8–12.1)	13.00±4.577 (12.1–13.9)	10.65±0.685 (9.8–11.5)

Contd...

Table 57–Contd...

Physico-chemical Parameters	Site–V	
	April to May	July to September
Temperature (°C)	25.40±2.753 (22.4–28.8)	25.83±0.976 (24.5–27.0)
Turbidity (NTU)	7.97±0.369 (7.5–8.5)	18.75±3.358 (14.5–17.5)
Velocity (cm/sec)	65.50±6.103 (58–75)	106.66±11.413 (90–125)
pH	7.60±0.122 (7.5–7.8)	7.73±0.376 (7.3–8.3)
Dissolved Oxygen (mg/l)	8.67±0.456 (8.0–9.2)	7.83±0.286 (7.5–8.3)
Hardness (mg/l)	167.50±5.590 (160–175)	112.66±11.430 (95–125)
Calcium (mg/l)	12.95±0.383 (12.5–13.5)	10.85±1.013 (9.7–11.8)

condition (Table 59 and Plates 21, 24 & 25). Some resulting fries were transferred to the river for stocking and ranching the seed in the natural riverine ecosystem.

Natural Breeding and Spawning Behaviour

Maturity Size and Age

Maturity size in the different ages of mahseers, *Tor tor* and *Tor putitora* were observed only after regular dissections of various sized mature male and female specimens during their breeding seasons. The males matured earlier than females in both the species. The matured specimen of both *Tor tor* and *Tor putitora* were observed the spawning seasons clearly marked in April to May and than in July to September. The main spawning period however were recorded from July to September but confirms only two sets of breeding period in mahseer species.

The smallest size of mature male of *Tor tor* was observed usually only after 242 mm in total length, *i.e.* after completion of 2+ years. Moreover, all males above 285 mm were found mature. The smallest sized mature females were attained a length of 320 mm. All females below 320 mm were invariably found immature and length groups

Table 58: Quantitative Assessment of Planktonic Population and Macro-benthic Density in Spawning Grounds of Mahseer in Ganga River and its Tributaries

Parameters	*Site–I*		*Site–II*	
	April to May	*July to September*	*April to May*	*July to September*
Zoolplankton (cells/ml.)	211.2±30.078 (180–260)	33.8±28.180 (08–76)	245.0±51.599 (180–305)	28.1±27.359 (08–82)
Phytoplankton (cells/ml.)	1002.5±311.4981 (690–1515)	290.6±79.658 (190–395)	1117.5±352.375 (730–1630)	355.0±112.516 (210–510)
Macro-benthos Density (ind. m^{-2})	19.2±7.327 (10–30)	6.0±4.753 (02–13)	28.7±12.557 (14–46)	7.0±5.932 (02–16)

Parameters	*Site–III*		*Site–IV*	
	April to May	*July to September*	*April to May*	*July to September*
Zoolplankton (cells/ml.)	177.5±43.084 (115–235)	29.8±30.818 (06–71)	176.2±21.614 (155–210)	23.8±23.819 (04–60)
Phytoplankton (cells/ml.)	1072.5±361.29 (680–1495)	271.3±67.473 (188–370)	1095.0±343.929 (740–1480)	267.5±59.476 (190–345)
Macro-benthos Density (ind. m^{-2})	27.5±8.906 (18–40)	7.3±5.750 (02–16)	19.5±8.645 (10–30)	6.8±4.996 (02–14)

Parameters	*Site–V*	
	April to May	*July to September*
Zoolplankton (cells/ml.)	212.5±21.650 (185–245)	33.6±29.937 (12–75)
Phytoplankton (cells/ml.)	962.5±340.339 (860–1545)	321.6±117.544 (170–470)
Macro-benthos Density (ind. m^{-2})	29.0±10.954 (17–45)	20.5±16.269 (6–48)

above 350 mm were found mature, *i.e.* within or after completion of 3 years, the age was calculated by scales and operculum studies. In *Tor putitora*, the males mature only after 260 mm in total length and 2+ years of age, while females of *Tor putitora* were attain the maturity only after 350 mm in total length and within or after completion of 3 years in age.

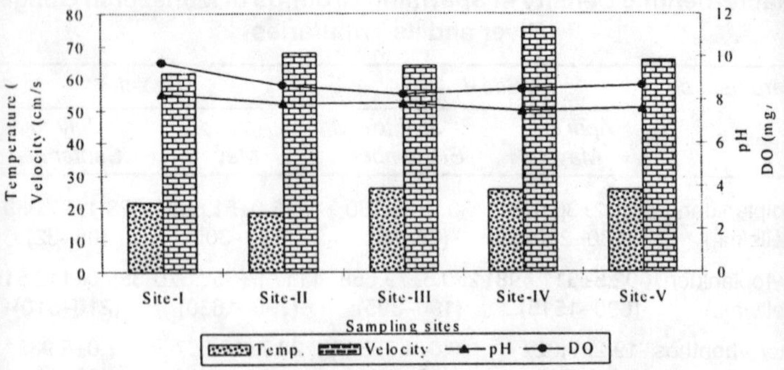

Figure 30(a): Physico-chemical Characteristic of Spawning Grounds of Mahseer in Ganga River and its Tributaries during April to May

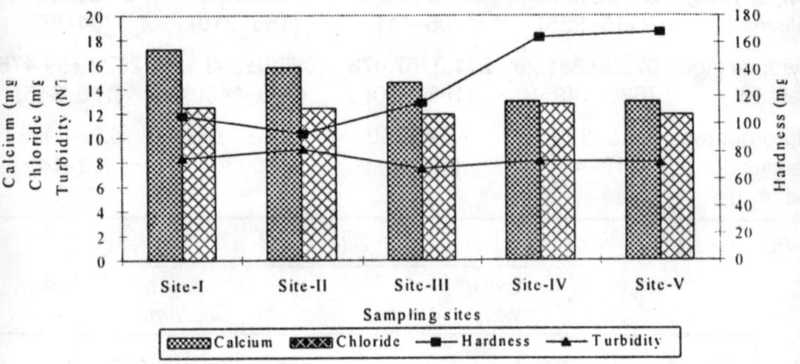

Figure 30(b): Physico-chemical Characteristic of Spawning Grounds of Mahseer in Ganga River and its Tributaries during April to May

Maturity Stages

The gonads of mahseer (*Tor tor* and *Tor putitora*) are paired elongated organs, suspended one on each side from the dorsal wall of the body cavity. They got progressively enlarged as the fish attained sexual maturity. The certain stages have been identified for males and females of the mahseer fishes separately as follows.

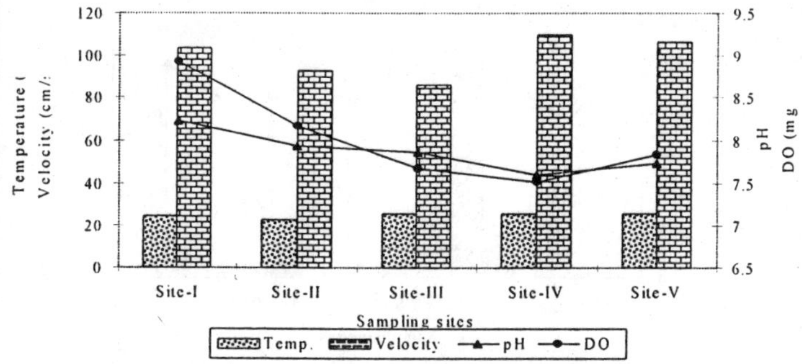

Figure 30(c): Physico-chemical Characteristic of Spawning Grounds of Mahseer in Ganga River and its Tributaries during July to September

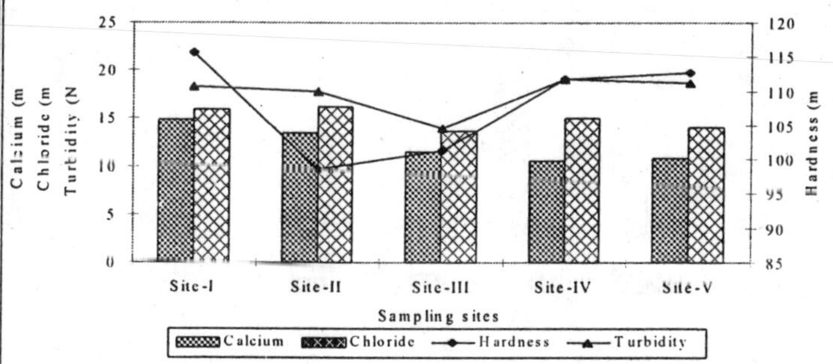

Figure 30(d): Physico-chemical Characteristic of Spawning Grounds of Mahseer in Ganga River and its Tributaries during July to September

Female Specimen

The microscopic examination of ovary at different stages of maturity was observed revealed four different stages based on oval size and maturity of gonadal structure.

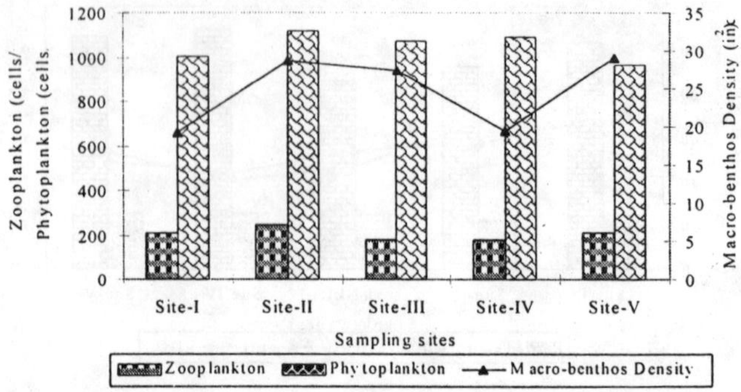

Figure 31(a): Planktonic Population and Macro-benthos Density of Spawning Grounds of Mahseer in Ganga River and its Tributaries during April to May

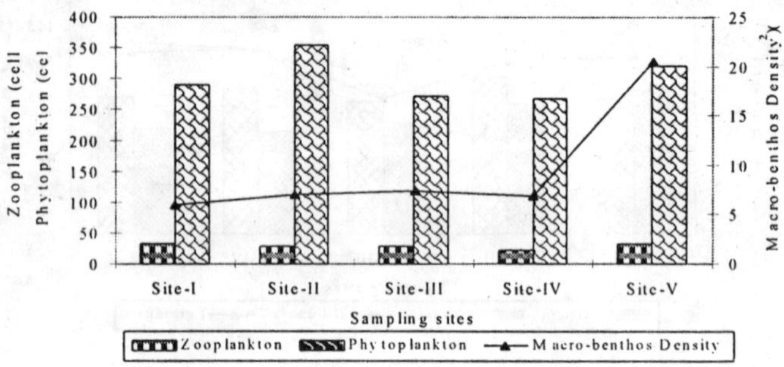

Figure 30(d): Planktonic Population and Macrobenthos Diversity of Spawning Grounds of Mahseer in Ganga River and its Tributaries during July to September

Stage 1

The ovaries remained small, thin, extending about half the length of body cavity, pink in colour, translucent, ova not visible by naked eyes and the relative weight of ovary below 0.8 to 0.9 per cent to body weight.

Table 59: Quantitative Survival Rate of Fertilized Eggs of Mahseer in Natural Aquatic System of Ganga River and its Tributaries (Under Controlled Conditions)

Period	*Ova Diameter Range (mm)*	*No. of Fertilized Eggs in Hatching Trays*					*Hatching Hours* Range	*Temperature Mean Value* (°C)	*%age of Mortality*	*Cumulative Survival Rate* %age
		Site–I	*Site–II*	*Site–III*	*Site–IV*	*Site–V*				
April–May	2.8–3.0	62	74	68	70	82	85.5–94.0	23.40	6.80	93.20
July–Sept	2.9–3.2	98	86	105	90	110	82.0–90.2	24.90	12.55	87.45

Stage 2

Length of ovaries extended about two third of body cavity, creamy white to yellowish in colour; ova opaque with average diameter ranging 1.96–2.05 mm. In advanced part of this stage, the ovaries even extended to the entire length of the body cavity with bright yellow, fully yolked, opaque or translucent ova having transparent periphery.

Stage 3

Ovaries formed up to sixteen per cent of the body weight. At this stage fully formed ova could be easily released by a slight pressure on ventro-lateral part of the abdominal cavity.

Stage 4

In this stage, fish possessed small and loose flaccid ovaries with a few remnants of ripe ova seen in the lumen.

Male Specimen

The following maturity stages were analysed in the male specimens of mahseer.

Stage 1

Small, thin, pinkish coloured testes were occupied about one-third of the body cavity and formed 0.5 per cent of the body weight.

Stage 2

The testes were observed as pinkish, opaque, translucent and fleshly opaque. The testes were thicker, elongated and extended to about half length of body cavity.

Stage 3

The testes were appeared enlarged, lobed and extended over the entire length of the body cavity; oozing milt on slight pressure on the abdomen during breeding season. Testes contributed 9.5 to 10 per cent of the total body weight.

Stage 4

The shrunken loose testes were recorded in a flabby condition extended to more than half of the length of the body cavity.

Spawning Habits

Spawning habit of mahseers (*Tor tor* and *Tor putitora*) was observed during early morning and evening hours in the selected spawning sites. Courtship is a long process, when several males chased the female at breeding sites. Two to seven males were seen closing one female, when the female found a suitable site, the eggs were laid and at the same time two to four males swam around the female contacting it by making rapid movements by the caudal region, after which they release their milt on the spawned eggs. After a short round movement by males and females returned towards downstream of the river. Which clearly shows that no parental care is taken either by male or female. The chasing and courtship period may be long but when the female lays eggs, the male pours the milt as soon as, the eggs are spawned and settled down on the stony pebbles substratum of spawning grounds or attached with adjoining aquatic vegetations.

The fertilized eggs, which were collected from spawning sites, were separated for *Tor tor* and *Tor putitora*. *Tor tor* has yellowish orange coloured and smaller sized eggs, while *Tor putitora* has lemon yellow or yellowish light orange coloured larger eggs. No colour differentiation of larvae of *Tor tor* and *Tor putitora* were observed as in the case of fertilized eggs.

Spawning Periodicity

Spawning period is an important aspect for the breeding biology of mahseer fishes and fishery management of a stock. Therefore, spawning season was made only after regular dissection of male and females of both *Tor tor* and *Tor putitora*. The ovaries and testes were also teased and examined under microscope during pre-spawning and spawning periods in order to determine their spawning season. Naked eye observations were also suitable to consider the spawning period due to large sized differentiated eggs in the ovary of the female, as also by eggs and milt oozing after slight pressure on the abdomen of the female and the males respectively.

The spawning season for *Tor tor* was remarked from April to September, when males yielded milt after slight pressure on the abdomen and the females exuded eggs by the same method. In October even a single female or male were not found gravid or mature. Microscopic observation of ovary and testes showed their spent condition. These observations have confirmed that the spawning season varied from April to September, which was also indicated by high condition factor both in male and females during this period. The eggs and larvae were also collected from May to October and a few fry and fingerlings were also caught after September onwards.

In *Tor putitora* repeated dissections of male and female fishes with similar technique as for *Tor tor* showed, the spawning season to vary from May to September. More males were observed to be mature in May and September than the females. The eggs and larvae of *Tor putitora* were available only May last to September, and few fingerlings were collected from June onwards only (Plates 21–25). The spawning season was also confirmed by high condition factor during this period. This revealed that mahseer laid the eggs in batches during the breeding season. The ova diameter range of mahseer were 2.8 to 3.0 mm and 2.9 to 3.2 mm recorded in April to May and July to September respectively, so it is an indication that the May and August are the peak months of breeding in both *Tor tor* and *Tor putitora*.

Measurements of Conservation Practice

On the basis of close ground field survey, lots of causative declining fishing methods were noticed during study period to catch the fishes in Ganga river. The main illegal fishing devices were used as follows:

Dynamiting

The worst explosive technique of dynamite blasting used in the running water of Ganga river. Such illegal practices were noticed in rainy and winter season nearby Phoolchatti, Garudchatti in upstream and Raiwala river zone in downstream of Ganga river (Plate 29).

Hammering

Hammering practices were used by the labour community, those are engaged in stone extraction activities in Ganga river. Such type of illegal practices were applied through diverting river water channels towards many directions, such type of practices have also observed in the different zones of the downstream at Raiwala and upstream near Saptsarovar.

Plate 28: Showing Deposition of Heavy Silt During Rainy Season in Downstream and Saptsarovar Region in Ganga River

Plate 29: Showing Dynamiting Activities for Fishing in Ganga River

Plate 30: Showing Electrocution Activites for Fising in Ganga River

Electric Shocks

The electric wire is passed to the stream and applied into a river pools and riffle segment of water. Such types of electrocution have noticed near Bengali residential clusters at Raiwala on the bank of river Ganga (Plate 30).

Bleaching Powder

The bleaching powder is commonly used for the purification of drinking water. It was used for the fish catching operation by the local fishermen in different natural pools or making the separate

Plate 31: Showing Cast and Gill Net's Operation for Fishing in Ganga River

Plate 32: Showing Underseized Mahseer Fishes caught from Ganga River

water channels in Ganga river. The technique of bleaching powder's application was noticed in two to four different segments of the main Ganga river and its tributaries.

Cast and Gill Nets

As far as the commercial fishery in the river Ganga, fishing is prohibited in upstream of Ganga and Saptsarovar at Hardwar region due to the religious nature of this town. But catch permits are issued to the contractors on tender basis for the different segments of Ganga river in Hardwar region. During study period, cast and gill net's operation were noticed in upstream of barrage and the confluence point of Song and Suswa tributaries, as well as confluence point of main Ganga river (Plates 31 & 32).

Ichthyotoxic Plant Extracts

Between Rishikesh-Hardwar region, many kinds of ichthyotoxic plants are used for fish catching in Ganga river system. Local people were used different parts of these plants in the pools and semi stagnant waters of Ganga river. The main ichthyotoxic plants and their parts were used in the fish catching practice are follows:

Name of the Plants	*Parts Used*
Khinna (*Sapium* species)	Leaves
Sulla (*Euphorbia royalana*)	Leaves
Tea (with lime)	Leaves
Trimal (*Xanthoxylum alatum*)	Pounded seeds
Mainu (*Randia dumentorium*)	Leaves
Akhrot (*Juglans regia*)	Leaves
Chilla (*Cascaria tormentosa*)	Pounded seeds
Rambans (*Agave* species)	Leaves
Khinkari (*Acacia* species)	Barks
Gahniri	Pounded seeds

Mahseer Migratory Routes

During the natural breeding process, migratory routes of mahseer species were also tried to found in the main Ganga river and its tributaries. Mahseer species occurs in the impoundment water of reservoir near the site of barrage. The bigger mahseer fishes of impoundment moves to the upper reaches towards Laxmanjhula, Garudchatti, Phoolchatti and lastly up to the Vyasghat and into the warmer waters of river Hanval and Nayar. While the young ones moves with the flow of Ganga river towards down side of Chilla Hydroelectric project and start their migration from the upstream of Bhimgoda barrage at Hardwar (Figure 29). First batch of ripe mahseer were noticed near the Chilla Hydroelectric power station, but they returned back into the main Ganga river due to blockage of upward migration route by the power project. The second batch of ripe mahseer moves into upstream towards Saptsarovar and reaches to downstream region of Pashulok barrage where minimum level of water remains found in the main river. Therefore, the mahseer fishes of Saptsarovar region migrate to the river Song and reaches towards Suswa river for breeding purposes. As a part of main conservation practice in kinds of *'in situ'* conservation, some quantity of the mahseer seeds as fingerling stages (Figure 26) were collected from the Vyasghat (upper reaches of the Ganga river) and transplanted into the downstream of Ganga river nearby Saptsarovar sampling station.

During the study period the main emphasis were given to create mass awareness among the local people, students, youth and Gram Pradhan or Sarpanch. Under this program, the mahseer conservation youth groups were created in five villages under the leadership of Gram Pradhan. These groups engaged to promote or restoring the natural fish habitat in the different fishing zones of main Ganga river and co-ordinate with forest department for conserve the mahseer in their areas.

Figure 32: Conservation Strategy for Mahseer Fishery Development

IMPACT OF BARRAGE (Main declining factors)

- Barrier effect (Population partition).
- Water quality changing process.
- Delinking of endemic fish population.
- Natural habitat loss in upstream and downstream.
- Paucity of drifts in food spectrum.
- Fish growth and maturity changes.
- Stoping mahseer migratory route.
- Silting of natural spawning/feeding sites.
- Water minimize/dried stages in downstream

ACTION ORIENTED PLAN

- Migratory channels in between upstream and downstream.
- Maintain miniman quantity of water for fishery in downstream.
- Construct seperate breeding channels along with main downstream river.
- Restoration of river system habitat, feeding and breeding spawning grounds.
- Seed stock enhancement
- Biomonitoring of the water quality.
- Aforestration on river basin catchment
- Formation of mahseer conservation youth groups in local area.
- Stop illegal fish catching practice.

FUTURE GOVT. ORIENTED PLAN

- Strictly implement modified fishery act.
- Ban in fishing in prebreeding and breeding season.
- Declare fish sancturies as Reserved Protected Area Network.
- Declare new fish conservation policy for threatened species.
- Co-ordination commitee of Central, State different departments.
- Action oriented groups from Central, State Universities and Govt. Research Institutes.
- Mass conservation action groups of local peoples and NGO.

Chapter 6

Discussion

In the development of fisheries in streams, the ecology of their surface runoff waters play the most important role since the ecology determines, the habitability and abundance of the flora and fauna in its different riverine sections. The impact of water obstruction by constructing the dam, barrages, hydroelectric power stations have shown marked changes in the water quality, biotic life and habitat degradation of coldwater fishes specially the mahseer, *Tor tor* and *Tor putitora* inhabiting in the upland water bodies.

A review of literature indicated that scanty information on the impact assessment of barrage and dams on the mahseer fish population dynamics, breeding behaviour and conservation measurement was available. No attempt has far been made to study about the impact assessment on bioecology of mahseer in Ganga river. Considering these aspects necessity of a comprehensive study on fish diversity, distribution and growth pattern, feeding habitats, natural breeding and spawning behaviours with special reference to conservation measures of the Himalayan mahseer were realized and will be great helpful to improve the present status of the mahseer population in Ganga river at Garhwal region.

Fish fauna occur in the Ganga river in the foothills of Garhwal Himalayas at Rishikesh as quite rich and diversified. There were thirty-five species belongs to nine families observed in Ganga river. Amongst the nine families, Cyprinidae have always dominant in Ganga river. A more or less similar occurrence trend have been recorded by Khanna and Badola (1990) and Khanna *et al.* (1998) in the Ganga river and Malik *et al.* (1998) in upland water of Himalayan region. Singh *et al.* (1983) reported that *Garra, Crossocheilus, Nemacheilus, Glyptothorax, Labeo, Puntius, Rasbora, Danio, Esomus, Mystus, Channa* and *Xenentodon* are common in Ganga river. Hora and Mukerjee (1936) and Om Kumar *et al.* (1990) also reported similar observations in Doon valley. In the present study, 3 species threatened, 9 vulnerable and 15 species rare categorized in Ganga river at Rishikesh region as also categorised by CAMP (1998). Jhingran (1990) also reported the population status of fresh water fishes in Ganga river in Himalayan region and mentioned one species was extinct, 5 were endangered and 22 were rare in Ganga river.

Several important reservoir systems in the country have been developed by construction of dams and barrages across the rivers for purposes of irrigation, flood control, navigation, generation of power etc. Dams, weirs and barrages act as physical barriers to migrations, tending to prevent access of the fish to their usual breeding, rearing and feeding grounds.

The Ganga river has a rich ichthyodiversity as 35 fish species belongs to 21 genera, 7 sub-families and 9 families. The Cyprinidae family occurs in a dominant phase in Ganga river. In the downstream, 12 species have recorded rarely showing negative effect on the fish population due to not maintained the minimum water requirement for fishery development in Ganga river. The deleterious effects of dams on fisheries in general have elucidated by Day (1873), Thomas, (1897), Hora (1940 and 1942), Chacko (1952 and 1954), Karamchandani *et al.* (1967), Karamchandani and Pisholkar (1967) and Kumar (1986). Dubey (1988) observed the low percentage of mahseer catche in total landing of fishes in Gandhisagar dam (M.P.). In the present study, percentage of mahseer catch was higher in the upstream of Ganga river while lower catch in downstream of barrage due to regulation and channelization of water by Chilla canal. Nautiyal and Lal (1981) and Sehgal (1988) have also reported that the percentage of mahseer catch reduced due to blocking of migratory passage and destruction of breeding grounds by dam and barrages in all hill-streams of Himalayan region.

David (1953) have conducted pre-impoundment studies on the bio-economics and some early stages of the Mahanadi mahseer,

Tor mosal mahanadicus, which showing a reduction in fish catches by 48.87 per cent within a period of fifteen years. The reasons for depletion of mahseer population can be attributed to indiscriminate killing during post monsoon and winter months when migration take place for breeding. On the contrary Valsangkar (1993) studied the fishery of mahseer in Koyna river after the construction of Shivaji Sagar reservoir and reported that the fish *Tor khudree* (Sykes) and *Tor mussullah* (Sykes) got adjusted to the changed ecological conditions and found its own ground for natural breeding. Similar observations were found in the Ganga river during the present study.

Nath *et al.* (1994) reported the Saung, Suswa, Tons, Giri, Asan and the confluences of these rivers with the Ganges and the Jamuna were particularly not worthy. Mahseer weighing upto 30 lbs were fairly common, 57 pound mahseer has been caught at Tejwala in the Jamuna river, while 60 to 70 lb fishes caught in the Ganges system recorded in the past. But in the present study maximum 1800 gms size of mahseer was caught in Ganga river, which indicated the declining feature of length and weight in mahseer of Ganga river. Sehgal (1971), Srivastava *et al.* (1985), Desai (1992) and Pathani (1994) have also reported that so many declining factors involved for drastic changes in mahseer fishery in the lentic and lotic water system of India.

The fish catch composition showed that the mahseer fishes have contributed about 58.00 per cent of the total catch in upstream during winter season and about 41.07 per cent of the total catch in downstream of Ganga river during the same period. The size group 461–500 mm of the mahseer as a dominant size of the catch have contributed 47.61 per cent but the same size group have also contributed 38.57 per cent of total mahseer catch in downstream of Ganga river. In Pong dam (Himachal Pradesh), the catch of *Tor putitora* contributed to about 5 per cent to 10 per cent in the year 1976–77 to 1979–80 and the mesh size for this mahseer was regulated to 30 cm. Srivastava *et al.* (1985) reported that three year class of *Tor putitora* having 380 mm size was found the most dominant in Govindsagar (being 95 per cent of the total catch of *Tor putitora*). According to Nautiyal and Lal (1984), mahseer ascend in upper reaches from the foothills for breeding purpose, which has also observed in present study at the early southwest monsoon when the brood mahseer migrate upstream waters upto the Song, Suswa and the Hanval river for breeding and spawning purposes. At the time of upstream migration of fishes, angler, fishermen and other people living in near villages hunting and killing the brood fishes in large quantities. The high percentage of mahseer catch in winter season have caught by the fisherman and anglers due to law water quantity in river stream, which has also supported by Pathani (1978), Joshi *et al.* (1993) and Nautiyal *et al.* (1998) and reported that maximum catch of *Schizothorax* in winter months in all Himalayan rivers.

In the present study, maximum catch (58 per cent) of mahseer has recorded in upstream of barrage but the percentage was found lowest (24.9 per cent) in downstream due to destruction of breeding and feeding grounds of mahseer, chocked by high siltation and sedimentation, high current of water, low water quantity and sometime partially or completely dried of river due to excess trapping of water for power generation. Similar observation have also studied by Srivastava *et al.* (1985), Sehgal (1994), Dhanze and Dhanze (1996) in other hill stream rivers of India.

In the present study, the size group of 191–350 mm fish forms the major part of the catche indicate that large number of fishes have landed before they could attain sexual maturity. The fishery pattern showed that the exploitation of the juveniles stocks of the Himalayan mahseer in the Ganga river and its tributaries have irrational. Similarly, Srivastava *et al.* (1985) reported that *Tor putitora* of 380 mm size was dominant in Govindsagar reservoir while Nautiyal *et al.* (1998) have bserved 310 mm size of mahseer dominant in Ganga and Song river.

The current status of the fishery in the Ganga river may be attributed to the over exploitative nature which off course is a consequence of extensive stream regulation. Moreover, there seems to be no control on the mesh size because all size of the fishes have recorded in the landing during the study period.

Biometric (morphometric and meristic) characters are useful for the identification of a fish species and detecting variations in fish population. Biometric studies reflect the proportionate growth of different body parts and the influences by environmental factors in a particular habitat. Intra specific variations among fishes may be due to ecological or genetic differences, which are important in fisheries as they provide rational basis of proper management and exploitation of fish population

According to Hubbs (1921), Vladykov (1934) and McHugh (1954) fish species were exhibit differences in their meristic and morphometric characters. Various characters like minimum and maximum body depth, eye diameter and preorbital distance have employed for the identification of fishes by Kirka (1974) and Livosvarsky and Ruban (1985).

During the present study, biometric characters have shown a proportional positive increase with increase in length of mahseer (*Tor tor* and *Tor putitora*) in both upstream and downstream of barrage. Total length and other body measurements show allometric growth. Alam (1977) observed morphometric characters like body weight, head length, depth of the body and diameter of the eye in relation to the total fish length. In *Rastrellinger kanagurata*, the head length and body depth grow simultaneously with the increase in length calculated by Mishra (1967) and Day (1878 & 1889). Similar observation have also been recorded by Iftekhar and Dwivedi (1977) in the morphometry of mahseer.

High degree of correlation have calculated for Standard length followed by Head length and least correlation for length of Rostral Barbel calculated in *Tor tor* of upstream. While, high regression coefficient for Standard length and least regression coefficient for length of Rostral Barbel have calculated with Total length of fish in downstream of the Ganga river. Singh and Dobriyal (1983), Dobriyal and Bahuguna (1987), Johal *et al.* (1994), Pandey *et al.* (1995), Pandey and Nautiyal (1997) and Singh and Singh (2000) have also observed high degree of correlation between different body parameters of various fishes and reported all body parts increased simultaneously with the increase of total length of fish.

During the present study, r-value of *Tor putitora* showed that snout length become progressively smaller in relation to total length and head length of fish in both upstream and downstream of the Ganga river. While, high regression coefficient for Standard length (0.811) and least regression coefficient for Eye diameter (0.001) were calculated with Total length of *Tor putitora* in downstream of Ganga river. Similar observation have been made by Tobar (1974) in *Lates niloticus* and by Dasgupta (1989) in *Acrossochelius hexagonolepis* of north-eastern region.

According to Bayagbona (1963), a constant index in any of the biometric characters in relation to its reference length is isometric. The growth of head length in relation to total length and growth of inter-orbital distance in relation to head length were recorded allometric growth in *Tor tor* and *Tor putitora* during the present study. Similar observation has been made by Dasgupta (1991), while according to Gould (1966), ratio between morphological characters are not necessarily constant for variation and difference in sex, race, nutrition and other environmental factors.

In the present study, standard length, head length, average height, length of pectoral fin and interobital distance were noted highly correlated with total length of *Tor putitora* in upstream. Maximum girth and interorbital distance were also highly correlated with total length in downstream of barrage. Interorbital distance was always highly correlated with head length in both upstream and downstream. Similarly above morphometeric characters were also highly correlated in *Tor tor* of upstream and downstream. But interorbital distance was least correlated with Total length while it was highly correlated with head length of *Tor tor* in upstream. The meristic characters were observed as lateral line scale and transverse scale in *Tor tor* and *Tor putitora.* According to Dasgupta (1989 and 1991), the morphometric characters of *A. hexagonolepic, T. tor* and *T. putitora* showed a proportional positive growth with the increasing total length and noted high degree of positive correlation with the reference length of fish in Simsang river at Garo Hills. Kirka (1974) reported the growth of some morphometric characters of *Salmo trutta m. fario* show significant differences from different elevations, while slope gradient and stream velocity show significant difference in *Thymalus thymalus* (Piivnicka and Hensel, 1978). According to them, lateral line scale, head length, length of dorsal fin, and length of pectoral fine are influenced by environmental factors and suitable for comparative biometric studies.

According to Hamilton (1822) and Hora (1940), the length of the head in proportion to the depth of the body is considered the most important character to distinguishing the mahseer species occurring in India. The mahseer *Tor tor* has characterising it as a form in which the head is considerably shorter than the depth of the body and the fins are reddish in colour, while *Tor putitora* has characterised by longer head than depth of body and the fins are tinge yellowish in colour. The colour varies according to the nature of water inhabited by the fish. Hora (1939) and Hamilton (1822) noted the colour as gold and green above, silvery below and the fins of the belly were reddish, while according to Hora (1940) the dorsal surface of the body was grayish green and head was neutral green. Similar observation were observed in present study and noted that head length was always shorter than body depth in *Tor tor* while head length was always greater in older specimen of *Tor putitora* than younger ones. Earlier, Day (1878) and later Mishra (1959) have also dealt with some similar biometric characteristics in mahseer. Lower lip with a median lobe of varying length. Two pair of barbels, maxillary barble longer than diameter of eye, reaching beyond posterior margin of eye, rostral barbel equal to sometime shorter than maxillary, similar observation have been made by Menon (1991) in *Tor putitora* in Daccan plateau.

Dasgupta (1989 & 1991) observed variation in morphometric and meristic characters of *A. hexagonolepis* in two rivers. Statistically significance difference only in case of the number of lateral line scale and concluded these variations have attributed to the difference in the environmental condition of the fish. Nautiyal (1983) and Nautiyal and Lal (1988) studied morphometric and meristic characters of *Tor putitora* and assess the heterogeneity in stocks of the river Alaknanda and rivulet Nayar in Garhwal region.

According to Dasgupta (1989 and 1991) the morphometric characters of *A. hexagonelepis, T. tor* and *T. putitora* have shown a proportional positive growth with the increasing length of the fish and noted high degree of positive correlation with the reference length in Simsang river. Results of present study indicate that the head of the fish grows faster than body itself, therefore it was concluded that the older specimens have longer head, than the younger ones. Correlation coefficients shows that corresponding growth of one part of the body with growth of the other, growth in lengths of head and standard length are associated with increase in total length, whereas the snout length least correlated with total length of mahseer. The similar growth pattern have been reported by Dubey and Dubey (1987) in mahseer, *Tor tor* of Narmada river.

Age and growth determining of wild population are much more difficult because they are more diverse, usually containing a broad assemblage of ages, life stages and sizes, all of which can differ by sex. Determination of age and growth of fishes have been described by Graham (1929) Chugunova (1959) and Casselman (1987). Fish growth is a continuous process of decreasing intensity, inversely proportional to age. Internal factors (spawning and physical stress) as well as external factors (temperature and food) acts upon growth causing cyclic variation in growth. This variation affects the development of calcified tissue during the growth phase in the fishes.

Recent investigations on scale and age of fishes have recognized four categories of age in the same year: 1 when annulus is at margin; 1+ when annulus near the margin; 1++ when the annulus is at some distance between last annulus and the margin; and 1+++ when distance between the last annulus and the margin is somewhat less than between two annuli. This pattern is also followed with average total length of scale and the average total length of fish for each year class in the present study.

Van Oosten (1929) and Casselman (1974) confirmed that there has a definite correlation between age and annuli in the scales, which increase as the fish grows from year to year. This relationship of annulus of scales and fish length were applied to *Tor tor* and *Tor putitora* in the present study.

Das (1959) has been demonstrated first time the feasibility of age determination by scales in tropical and temperate fresh water fishes of India. Annuli are formed mostly during winter season in cold or temperate waters, when scarcity of food in Ganga river (mainly downstream of barrage). A streak or crossing over of circuli takes place, which is evidenced in the scales and opercular bone of mahseer, collected during winter season. Raunsefell and Everhart (1953) have reported winter rings in fish scales due to low temperature as well as low feeding which was also observed in mahseer of Ganga river during the present study. The false rings were also observed in breeding season due to spawning acts of fish. Beckman (1942) also reported moreover false rings may be formed due to sudden environmental changes and fluctuations in water temperature or scarcity of food in a season.

Kagwade (1971) reported that the feeding intensity as the only factor for year mark in *Polynemus heptadactylus*, while no effect of spawning was observed in the fish scale. Jhingran (1977) stated that *Gudusia chapra* of the plains river has annulus formation during spawning season (April to July), when the fish has low feeding with low growth rate. According to Badapanda (1996) growth, maturation, gonadial stress and strain were reflected in the growth rings of scale in Mahanadi mahseer. Sinha *et al.* (1999) observed the growth is not significantly different from isometric growth in *Lepidocephalichthyes guntea* (Ham.). Tondon *et al.* (1993) have reported when the conditions are optimum and annual increment decreases with the increase in age. According to Sandhu *et al.* (1994) growth pattern of *Tor putitora* fluctuated due to under continuous stress in Govindsagar reservoir.

In the present study, the Scanning Electron Microscopic observation of mahseer scale exhibit that circuli show a regular pattern in the form of concentric rings and interrupted by the redii running in all directions. The marginal circuli have scaler denticles on the distal part and found also in the circuli break in annular zone during July-August and October. Johal *et al.* (1999) have also been studied similar interpretation in cycloid scale structure of *Tor putitora*. Khillare and Skora (1998) found hyaline zone in certain otolith due to cause of margin physiological stress, such as reproduction, metamorphosis, antagonistic migration or seasonal temperature change.

In the present study, the age determination by opercular method was found satisfactory even without confirmation of age by scale reading. These are many narrow circuli forming a broad annulus on the operculum. There were also false rings formed in two to five years old, as the fish mature after two years of age in both *tor* and *putitor* mahseers. These false rings indicate spawning and low feeding activity in a single year. The annuli identified as clear streaks, which have conspicuous over the broader part of the opercular (Lagler, 1959). The opercular bone was found satisfactory for back calculation and straight line relationship obtained between fish length and with of the bone, which have also done by Le Cren (1951) and Mann (1976) for fishes in British waters. It was also found that the fishes were less increase in length in higher age but increase in weight during the same age. Johal *et al.* (1999) have observed that back calculated length using opercular bone is slightly higher side compared to scales in *Tor putitora* from Govindsagar and noted that specific rate of linear growth show a decreasing trend with increase at age. Similar observations were noted in present study for mahseer in Ganga river. Hence the growth pattern showed that, the mahseer species appears to be under continuous stress in Ganga river.

The condition factor in the mahseer species were calculated from eviscerated weight eliminated the influence of gonads and food present in the alimentary canal. The pattern of fluctuation in these indices were almost similar to that of the indices obtained from the fish having gut and gonads, the difference being only in magnitude. These observations indicated that condition factor is not only influenced by the maturation of the gonads and the presence of food in the alimentary canal but also by the effect of seasons on the growth of body. The high condition factor (2.89–2.17) and lowest value (0.86–1.17) in summer and winter season respectively observed in *Tor tor*. As while in *Tor putitora,* the highest condition factor (1.155–1.069) and lowest (0.81–0.86) were observed rainy and winter season respectively in upstream and downstream of Ganga river. Similar observations have been made by Jhingran (1972) in *Setipinna phasa* and Nautiyal (1985) in *Tor putitora* in upper hillstreams of Himalayan zone.

The condition factor was also fluctuated due to high feeding in pre and post spawning period and combined with development in breeding season in both the mahseer species. Low condition factor were recorded during low feeding. But the condition factor and feeding intensity of *Tor tor* in Narbada (Desai 1970 & 1973), is the reverse of that for *Tor tor* in Ganga river. Nar Simham (1970) and Das and Moitra (1958) also studied fluctuating condition factor in some freshwater fishes of India. However, Kohlar (1960) have been reviewed the similar trends on condition factors of some fishes of south wales and Mraz (1964) described the similar observations in some fishes of USA waters. Lear (1969) showed monthly variation in condition factor in Greenland Halibul and Chonder (1973) reported the length-weight relationship in two blue gill populations in Rhode Island, where he concluded that weight increases more than length in fishes due to low density of fishes in the water body.

Mohan and Saraswat (2000) have observed increase in weight more than the cube of length in *Cyprinus carpio.* The relative condition factor (Kn) also found to be higher in small size fishes, which can be attributed to higher feeding rate. While, according to Dasgupta (1991), the average K-value highest during monsoon and lowest during summer. Nautiyal (1985) also reported similar trend in *Tor putitora* in Garhwal region. Such variations in the condition factor were attributed to factors like environmental conditions, food availability and gonadal maturity. The changes in the values of condition factor with the increase in length may yield evidences about the size at first maturity, while the seasonal fluctuation may reflect the condition of fitness and gonads (Jhingran, 1972 and Basimullah, 1975). During the present study, decrease in K-value with the increase in length of the fish indicated that juveniles have better condition factor. The Kn values were found maximum (1.069-1.155) in July and September due to full maturity of gonads and feeding in *Tor putitora* while this value decrease (0.862-0.810) in October and December followed "condition minus gonads". The high Kn values have shown the indicators of peak activity of gonads during July and September, which is also the peak breeding season of mahseer. Bali and Sharma (2002) have also been found similar observations for *Tor putitora* of Beas river in Distt. Kangra (H.P.).

Food and feeding habits of mahseer vary from one water body to another. Nikolsky (1963) recognised three main categories of food on the basis of their importance in the diet of fishes, they being:

1. Basic food, which is normally eaten by fish and covering most of the stomach contents.
2. Secondary food, which is frequently found in the stomach contents but in smaller amounts.
3. Incidental food, which is found rarely in the stomach contents.

Further, the fishes have again categorised as euryphagic, stenophagic and monophagic on the basis of variation in the food consumed.

In the present investigations, stomach of *Tor tor* and *Tor putitora* replaced by an intestinal bulb and the bile duct opens at the anterior third and middle of intestinal bulb respectively, which indicate the *Tor tor* characterised as herbi-omnivorous and *Tor putitora* was omnivorous, similar observations have also been reported by Lal (1968). Qualitative and quantitative seasonal food fluctuation of *Tor tor* and *Tor putitora* were showed that the fluctuation in response to the availability of food and its abundance in the environment of upstream and downstream of barrage and at Saptsarovar in Ganga river.

Nikolsky (1963) reported that abiotic factors regulating the food supply and affect the feeding habits of fish. In the upstream of barrage, more vegetative matter encountered than animal matter due to low temperature and other abiotic features due to dammed water of Ganga river in winter season, similar findings have been observed by Badola and Singh (1981), Dobriyal and Singh (1988) and Khanna *et al.* (1993, 1997 and 1998) in river water of Garhwal Himalaya. In downstream of barrage, water temperature, dissolved oxygen and pH of water were analysed higher than upstream due to high velocity of water. The increasing water temperature provided congenial conditions for the entomofauna to florish, which constitutes the basic food of *Tor putitora* and secondary food of *Tor tor*. Bisht and Das (1985) also reported that higher temperature is a favourable factor for an abundant and diversified entomofauna.

In Saptsarovar, the water temperature and other abiotic factors also highest but these factor were quite low than downstream of barrage due to confluences the greater amount of water of Song and Suswa river in Ganga river. So plant matter as well as animal matter always found abundantly in the river at this site. But in rainy season, the high turbidity and water velocity had a profound effect on the planktonic density by absorbing radiant solar energy in upper layer of water and thus impairing photosynthesis, which leads to a sharp fall in planktonic production. Joshi *et al.* (1996) and Khann *et al.* (1998) have analysed physico-chemical parameters of Ganga river in various stratches and observed that turbidity and vilocity affect the phytoplanktonic communities. Das and Srivastava (1959), Khanna *et al.* (1993) and Joshi *et al.* (1996) have pointed out that the water temperature play an important role as limiting factor for phytoplanktonic population in riverine ecosystem.

The analysis of gut content of *Tor tor* and *Tor putitora* were showed the animal feeding habit at young stages, which mainly feeds on insects and their larvae along with molluscs as the major part of their diet. Small quantity of macrophytes, algae (phytoplankton) and zooplankton were also encountered in feeding material of *Tor tor* and *Tor putitora*. Similar findings have been reported by Desai (1970) in juveniles of *Tor tor* from Narbada river and Pathani and Joshi (1980) and Malik (1994) in the fingerlings of *Tor tor* and *Tor putitora* in Bhimtal lake of Kumaon Himalaya.

In the present study, adult groups of *Tor tor* were showed herbi-omnivorous feeding habit, because plant matter (phytoplankton and macro vegetation) form the major part of their diet while in adult *Tor putitora,* more animal matter form the major part of their diet showed omnivorous feeding habit of fish. Hora and Mukerjee (1936) stated that *Tor* mahseer feed preferably on filamentous algae and water plants. Karamchandani *et al.* (1967), Desai (1970) and Pisholkar and Karamchandani (1981) have also reported that the fish *Tor tor* mainly herbivore and casually carnivore in nature. While, Das and Moitra (1955), Khanna and Pant (1965) and Badola and Singh (1980) have classified *Tor tor* as omnivorous and column feeder. Sharma (1987) identified the fish *Tor tor* carni-omnivorous in nature. However, during the present study *Tor tor* consumes more insects in its juvenile stage but thereafter, prefer mainly macrovegetation and other plants in upstream and downstream of barrage and at Saptsarovar sampling sites in winter season. But, *Tor putitora* shows the animal feeding habits in young or juvenile stages, they mainly feeds upon insect, crustaceans, worms along with molluscs, which form the major part of their diet. But in adult stages, the fish show omnivorous feeding habits due to more animal matter and less plant matter occurred as the major part of their diet. An appreciable quantity of sand and detritus was found in the gut of fish throughout the year, but in rainy season the quantity was much higher due to scarcity of basic food in the river by floods and silting in such a condition similar observation also reported by Nautiyal (1983) in fishes of the Alaknanda river in Garhwal Himalaya. Due to column and bottom feeding the sand, mud and debris were always found in the gut of hill stream fishes of Garhwal Himalaya described earlier by Badola *et al.* (1991). Thomos (1873) has also been reported that two pairs of barbels are indication of a bottom feeding nature of fishes.

The food and feeding analysis of *Tor tor* indicated the highest plant matter 62.24 per cent in the gut of *Tor tor* in upstream as compared 45.36 per cent plant matter in the gut of *Tor tor* occurs in downstream. In *Tor putitora,* the gut contained 50 per cent animal matter found in upstream and 42.79 per cent animal matter observed in gut of *Tor putitora* in downstream of Ganga river. The presence of plant and animal percentage were indicated that *Tor tor* preferred mostly plant matter and *Tor putitora* preferred mostly animal matter in upstream, downstream and Saptsarovar sampling stations of the Ganga river. High water temperature of downstream and Saptsarovar give suitable ecosystem to florish the entomofauna in these station which was also seen in the gut content of *Tor tor* and *Tor putitora.* Similar observations have been studied by Nautiyal and Lal (1984) and stated that *Tor putitora* mainly fed on aquatic insects, which occurred greatly in the ecosystem during water temperature recorded to higher at summer. Toor (1964) reported that the spectrum of different foods in riverine system which depends on many environmental factors and showed selective feeding in the fish, *Lethrinus lentjans* when both plant and animal food were available which is supported to the Saptsarovar site of the study. Frost (1977) also observed that correlation of the food intake with food availability in *Salvelinus neillighbi.* During the present study, plant animal ratio fluctuated seasonally, higher percentage of plant matter found in both the fishes during winter season when turbidity, water current and water temperature observed minimum, while high percentage of animal matter found in summer, when water temperature recorded maximum. It may be inferred that high temperature of summer is the favourable factor for abundance of insects in the river and provides an altering magnitudes of omnivorous to carnivorous feeding habit in *Tor putitora.* However, these types of variations in the feeding habit may be due to different nature of food organisms present in varying habitats of riverine ecosystem.

Qualitative and Quantitative studies of natural food of *Tor tor* made by Karamchandani *et al.* (1967) and Desai (1970) in river Namada while Dasgupta (1990) in river Simsang in Garo Hills. They have reported that the existing fish species were herbivorous, mainly feeds on higher aquatic plants, filamentous algae and also subsists on molluscs and insects. However, in the present study it was observed that *Tor tor* consumes more insects in its juvenile stage but thereafter mainly feed on macro-vegetation and other plant material in adult stage of life. Similarly, Jhingran (1975) reported *Tor tor* insectivorous in its juvenile stage but becomes herbivorous on reaching the adult stage, which is supported the results obtained in present investigation. The herbi-omnivorous nature of *Tor putitora* reported by Das and Pathani (1978) while food habits have shown a changing feeding pattern as omnivorous to carnivorous habits, exhibits an adaptation to the environment of the Nayar river reported by Nautiyal and Lal (1984).

Thomas (1973) observed mahseer feeds on small fish, while Desai (1970) encountered some bones and scales of large size fish in the gut of *Tor tor*. Pisholkar and Karamchandani (1981) have also been encountered flesh and long slender spines of fishes in the gut contents of *Tor tor*. Most of the authors considered fish as an incidental item in the diet of *Tor tor* and *Tor putitora*. However, in the present study, fish parts always found in the gut of *Tor putitora* while no fish matter found in gut of *Tor tor* except rainy season, when maximum food materials swept due to heavy rainfall, increased soil erosion and sweeping the bottom layer of Ganga river.

The observation of food volume, gastrosomatic index and feeding intensity were showed, that the mahseer of Ganga river feed actively during pre and post spawning periods when these values were found highest. But during spawning season, due to intermittent breeders of *Tor tor* and *Tor putitora*, they feed also intermittently as evidenced by low food volume, low gastrosomatic index and also by empty intestinal bulb during this period. The GSI value for *Tor tor* was observed highest (4.003) in winter and lowest (1.114) in rainy season in upstream and downstream respectively. While, the GSI value for *Tor putitora* was calculated highest (3.267) in winter and lowest (1.062) in rainy season in Saptsarovar and downstream respectively. Similarly, highest feeding intensity during winter and lowest during summer in *Tor tor* has been described by Desai (1970). Maximum feeding occurs during pre spawning due to maturation and enlargement of the gonads, while high feeding rate also noticed during post spawning period due to spent and starved condition of gonads in the mahseer fishes. But no seasonal feeding fluctuations have observed by Saigal (1964) in *Mystus aor*. Joshi and Kumar (1980) observed seasonal fluctuation in GSI of *Tor putitora* juveniles. Nautiyal (1984) and Nautiyal and Lal (1982) have also stated that GSI value maximum in winter and minimum in summer, while it was also lowest during rainy season (post spawning period) especially in fry and fingerlings of *Tor putitora*.

In the present study, the RLG values showed that *Tor putitora* is omnivorous, while a higher RLG value in *Tor tor* confirms the fish to be a herbi-omnivorous in nature. The RLG relationship was observed lower in fingerlings and higher in adult fishes of mahseer species, which is a reflection of their feeding habits. The low RLG value in *Tor putitora* indicated zoophagus or carnivorous feeding habits in the fingerlings or smaller specimen, while the high RLG values of *Tor tor* indicates it to be less zoophagus or herbiomnivorous in feeding habits. Khanna (1961), Jan and Das (1970) and Pathni and Joshi (1980) also recorded higher RLG values in herbivorous or in omnivorous fishes. Jhingran (1975) observed that the juvenile stages of *Tor* species have to be voracious feeders and prefers mainly animal matter in natural habitats.

Thus the general picture of RLG values were exhibit low RLG values in juveniles and higher in adults. This value also depends on the types of habitat and ecology of water body. Thus, RLG values were throw light on feeding habits in *Tor putitora* indicating it to be zoophagus in juvenile condition and less herbi and more carnivorous in adults stage, since more animal matter were found in the gut of *Tor putitora*. Therefore, *Tor putitora* categorized as carni-omnivorous in Ganga river system at Rishikesh. The contradiction thus arising due to the comparison of the actual dietary habits of the fish with the RLG value supports the view that the fish can adapt to the diet available in the particular environment (Steven, 1930; Pillay, 1953; Martin, 1954 and Singh, 1966).

Al Hussani (1947 & 1949) stated that purely herbivorous fishes have long gut while, Desai and Karamchandani (1968) observed the gut length ratio was less or equal than body length in carnivorous fishes. David *et al.* (1970) have studied the RLG value, and food and feeding habits, of *P. pulchellus* and concluded herbivorous in nature. Sehgal *et al.* (1984) reported that *Salmo trutta fario* have a short gut due to carnivorous feeding habit. Langer (1987) and Maheshwari *et al.* (1987) reported high RLG value in herbivorous and phytoplanktoniphagus fish *H. molitrix*. Pisholkar and Karamchandani (1981) observed *Tor tor* have long intestine and high RLG value because it feeds mostly on algal filament. In the present study, *Tor tor* have high RLG value while *Tor putitora* have low RLG value, therefore it was concluded that *Tor tor* herbivorous to herbi-omnivorous and *Tor putitora* was omnivorous in nature, which is also supported by Desai (1970) and Pisholkar and Karamchandani (1981). They have observedhigh RLG value and herbivorous nature of *Tor tor* in Narmada river.

Physiological changes are developed the gonadial development in fishes, which are responsible for sexual differentiation during breeding season in fishes (Hoar, 1957). Nikolsky (1963) also stated that development of secondary sexual characteristics have associated with the maturation of gonads in fishes. Chaudhary (1950) has reported that sex in some fishes could be differentiated only during their breeding season. According to Mishra (1950), the *Labeo boggut* male has transverse groove and pearl organs on the snout near the nostril. Similarly, Rao (1969) has been observed bright colour of fins and head in the male while dull colour in the female of band fishes (family-Copolidae).

The sex ratio of the mahseer fishes in Ganga river were recorded as the ratio of female and male of *Tor tor* were highest 1 : 3 in upstream and lowest 1 : 0.33 in Saptsarovar. The male specimen of the *Tor tor* were found maximum 66.03 per cent in upstream of Ganga river in rainy season. Whereas, the sex ratio of female and male *Tor putitora* observed 1 : 2.33 in upstream, 1 : 2.75 in downstream and 1 : 1.67 in Saptsarovar in Ganga river. The sex ratio of mahseer showed that more males were present in the Ganga river. Chaturvedi (1976) recorded the sex ratio of male and female to be 1 : 2 in *Tor tor* while Pathani (1978) observed sex ratio 1 : 1.33 for *Tor tor* and 1 : 1.56 for *Tor putitora* in Bhimtal lake. Similar sex ratio of female and male mahseer were found in the upstream and Saptsarovar site of Ganga river but the ratio was quite lower in downstream of Ganga river.

The sexual characteristics were also noted in *Tor tor* and *Tor putitora* in Ganga river during their breeding season. Both species were shows that the body colour, size and presence of tubercles in breeding season. The special characters which have seen in the fishes during their breeding season only. In this context the male were brighter in colour and smaller in size than female in both species *Tor tor* and *Tor putitora.* The pectoral fin was rough in males whereas it was smooth in females. Some tubercles were also present on the operculum near the snout, only in males of *Tor tor,* whereas no any tubercles were present in male specimens of *Tor putitora* during breeding season. Similar observations have also been reported by Lal and Chatrjee (1962) in *Tor tor* and *Tor putitora* in eastern Doon and Desai (1970) in Narmada river mahseer. Lips differentation as a character for sexual dimorphism also reported by Hora (1939 and 1940), Lal and Chartjee (1962) and Lal (1968) in *Tor tor* and *Tor putitora*. Kulkarni (1971) reported that male has comparatively longer and rough pectoral spine and some tubercles on operculum in *Tor khudree*. Khan and Jhingran (1975) have reported size differentation in *Labeo rohita* while George (1950) reported genital papilla in the cloaca of male *Sardinell longiceps* during breeding season.

All stream living mahseer species being starts their life cycles from eggs loosely buried in gravel on the streambed without making red or so called nest. The first preliminary account on spawning habits of mahseer were recorded by Day (1873), Beavan (1877), Nevil (1915) and Sken Dhu (1918), according to them mahseer deposit their spawn in several batches during a period of several months (May to August). But in present study, the breeding season extended from April to September for mahseers (*Tor tor* and *Tor putitora*) in river Ganga and its tributaries. Similar observation have been made by Thomas (1893) and Pathani (1982) and reported that the fish lays their eggs in batches and the pelvic and anal fins were used during spawning act by the male fish, which marks out a hollow in gravelly grounds. Khan (1939) recorded that mahseer spawn thrice in a year in rivers of Punjab while Smith (1947) reported that the mahseer breeds throughout the whole year. David (1953 & 1963) reported the spawning season of *Tor mosal mahanadicus* for November to January in Mahanadi river. Kulkarni (1971) recorded breeding season of *Tor khudree* from July to August, when high floods and low temperature occured in river. It seems that the start of rains or new waters from atmosphere has some bearing on the spawning season of the fish.

Himalayan mahseer spawn in such areas where the general substrate was gravel, rubbel and cobble (Shrestha, 1986) while, Jhingran and Sehgal (1978) reported sandy and gravely spawning grounds of mahseer in Himachal Pradesh. By the collection of some fertilized eggs in study period it was noted that the bed substratum of river gives suitable spawning substrate for the fish in breeding season. The bed substratum also gives a suitable live place for benthic fauna and planktonic community, which was feeding materials for fries and growing fingerlings.

The abiotic and biotic characteristics of the river water in spawning sites were indicated as a seasonal fluctuation trends. During the spawning season the water temperature 18.55-26.07°C, water velocity 61.50–110.33 cm/sec and dissolved oxygen 7.51–9.70 mg/l were analysed. The Zooplankton 23.83–245.166 nos./ml, Phytoplankton 267.50–117.50 nos./ml and macrobenthic density 6.0–29.0 ind/m^2 were also counted which provides suitable feeding substratum during the breeding. Khan (1945), Das and Dasgupta (1945) and Nautiyal (1984) have observed the thermal profile during the breeding season and recorded the range of temperature 21–25°C for breeding of Himalayan mahseer. Sehgal (1991) recorded hatching period for *Tor putitora* 80–96 hrs. in water temperature 22–24°C. Similar observations were recorded in throughout study period but sometime hatching period was high and low due to fluctuation of water temperature. An effort for natural developmental process of the collected fertilized eggs of the mahseer were tried in the shallow river water zone under control condition. A wooden hatching tray (size 60 × 30 × 10 cm) with fertilized eggs was kept in river water enclosed by happa. The incubation period 82–94 hours and mortality rate 6.8–12.5 per cent were observed at the water temperature 23.4–24.9°C. The total cumulative survival rate recorded 87.45–93.20 per cent at the stage of early swim-up fries and observed that the indigenous small flow in hatching trays should be operated to make more stocking maturity in natural condition for the mahseer fishery in Ganga river.

Mahseer prefers clean water and its migratory habits for breeding purposes observed by Cordingtion (1946). But due to rains the turbidity of water increased upto 19.13± 3.420 NTU because the sediments carried by floods and rainwater which adversly affect to bottom dwelling organisms. Ward *et al.* (1978) stated that sedimentation decreases substrate heterogenicity and fills intestines with silt which may severally reduce algal population and directly affect the zoobenthos. Therefore, the turbidity directly or indirectly affect the breeding of mahseer by reduce the feeding materials in spawning grounds. pH also affect the planktonic and benthic population, the maximum and minimum zoobenthic density coincided with maxima and minima of pH value. Ellis (1937) pointed out that aquatic organisms prefer pH between 6.7 and 8.4. During present study, pH value of spawning sites were slightly alkaline (7.1-8.8) while, Sundar and Joshi (1976) observed pH range 6.8–6.9 in spawning grounds of Angii stream.

Oxygen is an important constituent which incorporated the aquatic system as a result of photosynthesis and free atmospheric exchange at turbulence zone. Pathani (1982) observed high dissolved oxygen 7.5–11.5 ppm in water of spawning grounds of *Tor tor* and *Tor putitora* in Ramganga, Goriganga and Koshi river of Kumaun Himalaya. In the present study dissolved oxygen (7.5 to 9.7 mg/l) was analysed in spawning grounds of mahseer in river Ganga and its tributaries in Rishikesh-Hardwar region. The increased current velocity and turbidity during rainy months, washes plankton from the spawning grounds and minimum occur during the spawning period. The water velocity recorded 50-125 cm/sec in spawning grounds of mahseer during the study period while, Shrestha (1986) has been recorded velocity over the spawning ground 12–18 inches/sec. Joshi (1988) observed the velocity 0.12–0.16 m/sec of spawning grounds in throughout study period. Some other factors like hardness, calcium, magnesium and chlorides also affect the hatching of eggs, however no direct effect on benthic communities but affected planktonic biomass directly. Shrestha (1994) have also observed dissolved solids in water of spawning grounds of *Tor tor* and *Tor putitora* in Himalayan waters of Nepal.

The abiotic and biotic environments are interdependent, any changes in one system invariably produces changes in the other because of this subjective inter-relationship of fishes with abiotic environment and other organism. During the study period, the spawning grounds constitute sandy bottom, pebbles creeks of large stones and aquatic weeds with high water temperature and high dissolved oxygen. The fertilized eggs and fry of mahseers (*Tor tor* and *Tor putitora*) were collected along sand and pebbles at bottom in spawning sites which shows the eggs are lithopelagic similarly, Hannook *et al.* (1976) have been reported the spawning behaviour of barbus in shallow gravelly and sandy water of streams in British waters.

A trend of multiple spawning were observed in mahseers (*Tor tor* and *Tor putitora*) in the Ganga river. The observation revealed that the mahseer fishes spawned in two seasons dominantly during April-May as the first set, while the second set being in July-September. The similar observation pattern have also been described by Dunsford (1911), Khan (1939) and Sehgal (1974 and 1992). In the present study, the regular field observations were showed that the mahseer fishes laying their eggs in two phase as in April-May and secondly in July-September indicating two main breeding season that is pre-summer and rainy season.

The size of the collected fertilized eggs from spawning sites were recorded 2.80–3.0 mm in April-May and 2.90–3.2 mm in rainy season. According to Pathani (1980), the diameter of fertilized eggs ranged from 2–3 mm. According to Bhatnagar *et al.* (1977) depth of spawning ground of *Tor tor* was 1.50 m where 4 to 7 fishes spawned in Govindsagar, but in the present study the breeding grounds were observed shallow region of river with the depth of 35 to 50 cm having pabbly and gravely booom. Jhingran and Sehgal (1978) have reported deeper spawning grounds with sand and gravel of *tor* and *putitor* mahseers in Himachal Pradesh.

Dams, barrages and weirs constructed along the major streams and rivers of the country were affected the mahseer of India (Pathani, 1982). Dam construction, large scale collection of gravel, stones and silting of rivers and reservoirs affect the mahseer fishery. So it is necessary to conserve the natural habitats of mahseer by using different conservation measurements. Conservation is the management of the natural resources and controlled harvest with replacement of stock and maintenance of the habitat.

All dams must have inbuilt fish ladders and fish ways, which must be regularly maintained and to allow fish to migrate their spawning grounds in upstream or upper reaches of Himalayan region. The inbuilt separate running fish channel must be created in the vicinity of the impounding of the waters as a result of construction of dam or barrages should be utilized to maintain the healthy brood stock of the endangered or threatened fishes of the native area. Strict inforcement to prevent poaching and destruction of fishes by such illegal methods as use of explosive, electrocution, use of small mesh size's net, poisoned etc. The fishery and forest staff must be supplemented by appointment of honorary wild life fishery wardens who may be preferably selected from exservicemen or retired government servants. In this context, spawning grounds, brood fish and juveniles must be give special protections under the restoration programme to develop natural habitat for mahseer species in Ganga river system.

The fishery of the existing and proposed dams, barrages and reservoirs should be carefully planned and develop as an integral part of the riverine ecosystem. This will be possible when all hydroelectric projects in the uplands are properly planned jointly by engineers, fishery scientist, geologist, sociologists, ecologist and social workers. Provision for suitable fish passes or fish ladders should be made in all the proposed dams.

It is essential to prevent the steady decline of the world famous Himalayan mahseer by devising all possible methods of conservation, management and rehabilitation. Before implementing any steps on these lines, the various depleting factors as discussed above need to be removed at the first phase. The irreparable harm caused to the mahseer fish habitat need to be compensated through adoption of improvement measures like afforestation, eco-restoration, soil conservation efforts, completation on deforestation in sloppy areas and strictly implementation of *Indian Fisheries Act, 1897* (modified, 1956). Declare mahseer sanctuaries in selected protected riverine sources along with to make a new fish conservation policy for threatened species. All conservations measures would positively help to restoration of the Himalayan mahseer in Ganga river.

Chapter 7

Summary

The present book mainly deals with physio-geographical features, meteorological parameters and impact assessment of barrage on the mahseer fishery inhabited in the Ganga river, specially emphasizing population dynamics, breeding and spawning behaviour including the conservation measures of the mahseer, *Tor tor* and *Tor putitora*. The main findings of these investigations are summarized as follows:

1. Rishikesh, a pilgrim township of district Dehradun (Uttaranchal) is situated in the foothills of Shiwalik Himalaya on latitude 30°06′ N and longitude 78° 19′ E, on the bank of main Ganga river. The height of the town is 356 above msl consisting the surrounding area of the town is approximately 11.20 sq. kilometer.
2. The Pashulok barrage has constructed across the main Ganga river dividing as upstream and downstream river basin. The sampling stations were selected in upstream, downstream and Saptsarovar for analysing the various parameters during the study period.
3. The impact of meteorological observations were showed clear seasonal variations in different parameters *i.e.*, atmospheric and water temperature, relative humidity and rainfall. The climate of Rishikesh and its surrounding riverine zone was observed as more colder than plains and can be categorized as sub-tropical.

4. The Ganga river has a very diversified fish biodiversity as 35 fish species belongs to 21 genera, 7 sub-family and 9 families. The Cyprinidae family occurs in a dominating phase. In the downstream, 12 species were recorded rarely showing negative effect of barrage on the endemic fish population due to not maintained the minimum water requirement for fishery development and work as a physical barrier to segregate the inhabited natural fish species.
5. The fish catch composition showed that the mahseer fishes were contributed about 58 per cent part of the total catch in upstream during winter season and about 17 per cent mahseer catch have decline in downstream during the sam eperiod.The size group 461–500 mm of the mahseer as a dominant size of the catch have also contributed 47.61 per cent in upstream but the same size group have again shown 9 per cent declining pattern in downstream of Ganga river.
6. The biometric analysis of the mahseer species were showed that the body parameters, standard length and head length were highly correlated but snout length and length of the rostral barbel were least correlated with total length of the fish *Tor tor* and *Tor putitora* in both upstream and downstream of Ganga river.
7. In *Tor tor,* the total length 30.5±3.628 and head length 7.22±0.458 were recorded and high regression coefficient were calculated for SL (0.8276) and least (0.0004) for LRB found in the upstream. Average head length (6.56) and average snout length (3.08) of 33.9 cm average size group of fishes. High degree of correlation of SL (0.9906) followed by HL (0.9515) and least correlation (0.7095) of SnL with TL were calculated in downstream.
8. In the *Tor putitora,* total length 40.125±8.002 and head length 9.55±1.709 were recorded and high regression coefficient were calculated for SL (0.8828) and least (0.0069) for average Height found in the upstream. Average head length (7.72) and average snout length (4.21) of 34.00 cm average size group of the fishes, high degree of correlation of SL (0.9972) followed by HL (0.9929) and

least correlation (0.6110) of SnL with TL were calculated in downstream.

9. The meristic characters were observed as lateral line scale and transverse scale in both the species of mahseer. The lateral line scale counts have observed some differentiation in number as 24–28 in *Tor putitora* and 20–27 in *Tor tor.* Two pairs of rostrals barbels and maxillary barbels were found in the species of mahseer.
10. The age and growth of the mahseer species were calculated by scale and operculum method. False rings were recorded in scales and operculum due to low feeding rate in spawning period. Only two year age group and above than two year have clear false rings in scales and operculum of the mahseer, *Tor tor* and *Tor putitora.* The back calculation showed a linear relationship between total length of fish and scale radii along with the total length of fish and operculum width. By the present observations a separate formula for age determination given for *Tor tor* and *Tor putitora* separately for upstream and downstream of Ganga river.
11. The condition factors as standard formula $K = W \times 10^5/L^3$ indicated high condition factor (2.89–2.17) and lowest value (0.86–1.17) in summer and winter season respectively observed in *Tor tor.* As while in *Tor putitora,* the highest condition factor 1.155 and lowest value 0.81 were observed in rainy and winter season respectively. The condition factors were highest in breeding season and its value further decrease in winters (post spawning period) followed by condition minus gonads. The values of K have shown as an indicators of the peak activity of the gonads during the summer and rainy seasons, marks as a separate breeding seasons in two peaks for the mahseer species, *Tor tor* and *Tor putitora.*
12. The food and feeding analysis of *Tor tor* were indicated that the highest plant matter 62.24 per cent in the gut of *Tor tor* occured in upstream as compared with the value of plant matter 45.36 per cent in the gut of *Tor tor* occured in downstream. In *Tor putitora,* the gut contained 50 per cent

animal matter in upstream and the gut of *Tor putitora*, 42.79 per cent animal matter occured in the downstream. The presence of plant and animal matter percentage have indicated that *Tor putitora* preferred mostly animal matter and *Tor tor* preferred mostly plant matter in upstream, downstream and Saptsarovar sampling stations of the Ganga river.

13. The RLG values for *Tor tor* were recorded 2.397–2.293 in upstream, 1.840–1.754 in downstream and 1.915–1.855 in Saptsarovar region collected. The RLG values were found 2.129 in upstream, 1.562 in downstream and 1.865 were observed in Saptsarovar region specimens of *Tor putitora*. These value showed that *Tor tor* has high RLG value which indicated herbivorous or herbiomnivorous feeding habit, while the low RLG value of *Tor putitora* indicated omnivorous or carniomnivorous feeding nature of this species in Ganga river.
14. The GSI values of *Tor tor* were calculated highest (4.003) in winter and lowest (1.114) in rainy season in upstream and downstream respectively. The GSI values of *Tor putitora* were calculated highest 3.267 and lowest 1.062 in winter season at Saptsarovar and downstream of Ganga river respectively.
15. The sex ratio of the mahseer fishes in Ganga river have shown as the ratio of female and male of *Tor tor* 1 : 3 in upstream, 1:1.67 in downstream and 1 : 0.33 in Saptsarovar. The male specimens of the *Tor tor* were found 66.03 per cent in upstream in rainy season. The sex ratio of female and male of *Tor putitora* were observed 1 : 2.33 in upstream, 1 : 2.75 in downstream and 1 : 1.67 in Saptsarovar. The male specimen of *Tor putitora* were found 65.83 per cent in upstream during rainy season.
16. The physico-chemical parameters and biotic characteristics of river water of the spawning site have indicated as a seasonal fluctuation trends. During spawning season the water temperature range 18.55–26.07°C, water velocity ranges 61.50–110.33 cm/sec and dissolved oxygen's value 7.51–9.70 mg/l were observed,

while the zooplankton (23.83–245.166 nos./ml), phytoplanktons (267.50–1117.50 nos./ml) and macrobentic density (6.0–29.0 ind/m^2) were calculated in the spawning sites of mahseer. These characters of water showed that the mahseer fishes preferred more clear water, low water temperature, relatively high oxygen value and good feeding substratum during the breeding seasons.

17. The regular field observations have shown that the mahseer fishes laying their eggs in phases as in April-May and secondly in July-September which indicating two main breeding seasons i.e. pre summer and rainy season. The size of the collected fertilized eggs from the different spawning sites were recorded 2.8–3.0 mm in April–May and 2.9–3.2 mm in rainy months of July-September.
18. Spawning habits of the mahseer were observed during early morning and evening hours on the selected spawning sites as courtship, chasing by male specimens and rapid movement by the caudal region of the female during the spawning process. No parental care is taken either by male or female specimen of the mahseer.
19. An effort for natural developmental process of the collected fertilized eggs of the mahseer were tried in the shallow river water zone under control condition. A wooden hatching tray with fertilized eggs was kept in river water enclosed by happa. The mortality rate 6.8–12.5 per cent and incubation period 82–94 hours was observed at the water temperature 23.4–24.9°C. The total cumulative survival rate recorded 87.45–93.20 per cent at the stage of early swim-up fries and observed that the indigenous small flow in hatching trays should be operated to make more stocking maturity in natural condition for the mahseer fishery in Ganga river.
20. During the study period, weekly river shoreline survey work have shown a lot of destructive fishing methods were found operated by local fisherman, anglers and local labourers in the main Ganga river and its tributaries. These methods are dynamiting, hammering, electric shocks, bleaching powder, cast and gill nets and use of Ichthyotoxic plant extract to catch maximum fishes.

21. In conservation measure practice, the restoration of the spawning sites were made the optimum natural conditions for breeding of mahseer. After rainy season, heavy siltations in the Ganga river basin catchment destroy the spawning and feeding grounds. The unscientific extraction of stones, pebbles, gravels and sand from the Ganga river and its tributaries also affect the breeding and spawning habit of the mahseer.
22. Under *'in situ'* conservation program, approximately 1500 fingerlings (50–60 mm length) of mahseer were collected from Vyasghat and transplanted in Saptsarovar region of Ganga river.
23. During the study period, the mahseer conservation youth groups in the leadership of local village Pradhan and Sarpanch were also developed. These youth groups have really provided a great help to closely monitor the destructive fishing and other activities in the Ganga river. During the pre-breeding and breeding season, the mahseer conservation youth groups have paid the full attention towards restoration of the natural breeding habitats in all adjoining river tributaries and main Ganga river in Rishikesh region of the Himalayan zone.

At last, the central and state government should be more emphasized on the mahseer fish conservation oriented plan and implement strictly modified fisheries act, declare fish sanctuaries as reserved protected area network and declare new fish conservation policy for the threatened endemic fish species occurs in upland aquatic resources of Himalayan region.

Bibliography

A..G. Jhingran (1977). Optical appearance and interrelation of annuli on the scales of *Gudusia chapra* (Ham.). *J. Inland Fish Soc. India,* **9**: 138–153.

Alam, S.K. (1977). Morphometric studies of a Indian Mackerel, *Rastrellinger kanagurta* (Cuvier, 1817) of Bombay region. *Bulletin of the Central Institute of Fisheries Education:* 199–201.

Al-Hussaini, A.H. (1947). The anatomy and histology of the alimentary canal of the plankton feeder *Atherina forskal* (Rupp.). *J. Morph.,* **79**: 251–286.

Al-Hussaini, A.H. (1949). On the functional morphology of the alimentary tract of some fishes in relation to difference in their feeding habits, Anatomy History. *J. Mar. Sci.,* **92**: 190–240.

Alikunhi, K.H. (1952). On the food of young carp fry. *J. Zool. Soc. India,* **4(2)**:77–78.

Allen, K.R. (1935). The food and migration of the perch, (*Perca fluviatilis*) in Himdemere. *J. Anim Ecol.,* **4**: 264–273.

Annon, (1992–93). *National Bureau of Fish Genetic Resources,* Annual Report, p. 65.

Annon, (1993-94). *National Bureau of Fish Genetic Resources,* Annual Report, p 71.

APHA (1995) *Standard methods for examination of water and waste water.* American Public Health Association. 19th edition, p. 1750.

Badapanda, H.S. (1996). The fishery and biology of Mahanadi mahseer *Tor mosal mahanadicus* (David). *Indian J. Fish.*, **43(4)**: 325–331.

Badola, S.P and Pant, M.C. (1973). Fish fauna of the Garhwal hills, Part–I. *Ind. J. Zool.*, **14(1)**: 37–44.

Badola, S.P. and Singh, H.R. (1977): Fish fauna of the Garhwal hills, Part–IV. *Ind. J. Zool.*, **18(2)**: 155–168.

Badola, S.P. and H.R. Singh (1980a). Spawning of some important coldwater fishes of the Garhwal Himalaya. *J. Bombay Nat. Hist. Soc.*, **81**: 54–58.

Badola, S.P. and Singh, H.R. (1980b). Food and feeding habits of fishes of the genera *Tor, Puntius* and *Barilius*. *Proc. Indian Nat. Sci. Acad.*, **346**: 58–62.

Badola, S.P. and Singh, H.R. (1981). Hydrobiology of the river Alaknanda of Garhwal Himalaya. *Indian J. Ecol.*, **8(2)**: 269–276.

Badola, S.P. and Smita Badola (1999). Observation on destruction of fishing methods and recommendations to protect fish fauna in Garhwal Himalaya. In: Khanna, D.R.; Ashutosh Gautam and Aditya Gautam (eds.) *Sustainble Ecological System and Environment*, pp.165-168.

Badola, S.P., H.R. Singh and D.R. Khanna (1991). Food and feeding habits of some hill stream fishes of Garhwal Himalaya. *Him. J. Env. Zool.*, **5**: 141–143.

Bali, R.K. and K.B. Sharma (2002). Length-weight relationship and relative condition factor of *Tor putitora* (Hamilton) inhabiting Beas river in district Kangra. In: P. Das; J.R. Dhanze; S.R. Verma and D.S. Malik (eds.) *Coldwater Fish Genetic Resources and their Conservation*. Natcon Pub., 7: 195–202.

Barrington, E.J.W. (1957). The alimentary canal and digestion. In: *"The Physiology of Fishes"*. (M.E. Brown, ed.) Vol. 1. Academic Press, New York and London. p. 785.

Basimullah, A.K.N. (1975). Biology of *Lutijanus griseus* (L.) of the Cubaguna Island Venezuela L. length-weight, body length-gut length relationship and condition factor. *Biological Institute of Oceanography University Oriento*, **14(1)**: 101–107.

Bayagbona, E.O. (1963). Biometric study of two species of *Psycudotalithus* from the Lagos trawling ground. *Bulletin de Ifan T* **XXV** Ser. A1.

Beaven, R. (1877). *Handbook of freshwater fishes of India.* p321.

Beckman, W.C. (1942). Annual formation in the scales of certain Michigan game fishes. *Pap Mich. Acad. Sci.,* **28** : 281-312.

Berg, R. (1979). External morphology of the pink salmon, *Oncorhynchus gorbusha* introduced into lake superior. *J. Fish Res. Board. Can.,* **36 (10)** : 1283-1287.

Bhatnagar, G.K. (1964). Spawning and fecundity of Bhakra reservoir fishes. *Indian J. Fish.,* **11(1)** : 485-502.

Bhatnagar, G.K. and S.J. Karamchandani (1970). Food and feeding habits of *Labeo fimbriatus* (Block) in river Narbada near Hoshangabad (M.P.). *J. Inland Fish. Soc. India,* **2**: 30-50.

Bhatnagar, G.K.; S.J. Karamchandani and M.D. Pisholkar (1977). Observation on the spawning of major carps of Govindsagar reservoir, *Tor tor* (Hamilton) and *Tor putitora* (Hamilton). *J. Bombay Nat. Hist. Soc.,* **79** : 525-530.

Bhatt, J.P.; Nautiyal, P. and Singh, H.R. (1998). Comparative study of morphometric characters of Himalayan mahseer, *Tor putitora* (Ham.) between Ganga and Govindsagar reservoir stocks. *Indian J. Fish.,* **45(1)** . 85-87.

Bhaumik, U. and Saha, S.K. (1998). Role of extension in arousing mass awareness and public participation in fish conservation movement. *In* : A. G. Ponniah; P. Das and S. R.Verma (eds.) *Fish Genetic and Biodiversity Conservation.* Natcon Pub. No.5: 157-163.

Bilton, H.T. and D.W. Jenkinson (1968). Comparison of the otolith and scale methods for aging Sockage, *Onchorhynchus nerka* and *Onchorhynchus chumi. J. Fish Res. Bd. Canada,* **25**: 1067–1069.

Bisht, R.S. and S.M. Das (1985). Abiotic parameters and aquatic insects of two cold water Kumaon lakes. *Indian. J. Ecol.,* **12(2)**: 365–369.

Blake, C. and Blake, B.F. (1978). The use of operculam bones in the study of age and growth in *Labeo senegalensis* from lake Kainji Higeria. *J. Fish Bio.,* **13**: 287–295.

Boujard, T. (1995) Dial rhythms of feeding activity in European catfish, *Silurus glanis. Physiol. Behav.*, **58**: 641–645.

CAMP (1998). *Conservation Assessment Management Plan.* Conservation assessment and management plan workshop report, p. 156.

Carlander, K.D. (1953). Use of gill-nets in studying fish populations of clear lake, Iowa. *Proc. Iowa Acad. Sci.*, **60**: 623–625.

Casselman, J.M. (1984). Determination of age and growth. In: *"The biology of fish growth."* Eds. A.H. Weatherely and H.S. Gill, Academic Press, London: 209–242.

Casselman, J.M. (1987). Analysis of hard tissues of pike *E. sox lurus* with special reference to age and growth. In: *"The Ageing of Fish"*. Ed. T.B. Bagenal, Unwin Brothers Ltd., Surrey, England. 13–27.

Chako, P.D. (1952). Report on a survey of the dams and migratory fishes of Madras. *Cont. Freshwater Fish Biology Sin.* pp. 41–81.

Chako, P.D. (1954). Past, present and future of the hilsa fisheries in Madras state. *J. Asiat. Soc.*, **20(1)**: 55–58.

Chaphekar, S.B. and Mhatre, G.N. (1983) : A report of the project "*Impact of Human settlements and developement Activities on the Ganga river system*". Institute of Science, Madain Cama Road, Bombay. 65p.

Chaturvedi, S. K. (1976). Spawning biology of mahseer *Tor tor* (Ham.). *J. Bombay Nat. Hist. Soc.*, **73(1)**: 63-73.

Chaudhary, H. (1959). Notes on external characters distinguished sex of breeders of the common Indian carps. *Sci. and Cult.*, **25(4)**: 258-259.

Chonder, S.L. (1973). A possible seperation of races of *Gudusia chapra* (Hamilton) by mean of length-weight correlation. *Proc. Indian Acad. Sci.*, **68 (2)**: 73-79.

Chugunova, N.I. (1959). "*Age and growth studies in fish*". A systematic guide for Ichthyologists Translated from Russian Published for the National Science Foundation, Washington D.C. by the Israel programme for scientific translations.1963p.

Chugunova, N.I. (1963). "*Handbook for the study of age and growth of fishes.*" National Science Foundation, Washington.132p.

Coad, B.W. and Power, G. (1974). Meristic variation in the three species stickleback *Gasterotesus aculeatus* in the Matamels river system Quebec. *J. Fish Res. Bd. Canada*, **31** : 1155-1157.

Cordington, K.De.B. (1939-46). Notes on Indian mahseers. *J. Bombay Nat. Hist. Soc.*, **46** : 336-344.

Dahl, Lea (1910). *The age and growth of salmon and trout in Norway shown by their scales.* The Salmon and Trout Assoc. pp1-144.

Das, S.M. and Moitra, S.K. (1955). Studies on the food of some common fishes of Uttar Pradesh, India (part-I). *Proc. Nat. Acad. Sci. India*, **25 (I&II)**: 1–6.

Das, S.M. and S.S. Pathani (1978a). A study of the affect of lake ecology on the productivity of mahseer, (*Tor tor* and *Tor putitora*) in Kumaon lakes, India. *Matsya*, **4**: 25–31.

Das, S.M. and Pathani, S.S. (1978b). Studies on the biology of Kumaon mahseer *Tor putitora. Indian J. Anim. Sci.*, **48(6)**: 461–465.

Das, K.N. and Dasgupta M. (1945). Breeding of the principal carps in Bengal symposium on the factors influencing the spawning of Indian carps. *Proc. Nat. Inst. Sci. India*, **11**: 324–327.

Das, P. (1988). Conservation and Management of Fish Genetic Resources of India. In: Arun, G. Jhingaran and V. V. Sugunan, Ed. *Conservation and Management of Inland Capture Fishery Resources*, pp. 20–24.

Das, P. (1994). Strategies for conserving threatened fisheries. In: Dehadrai, P.V.; P. Das and S.R.Verma (eds.). *Threatened Fishes of India.* Pub. No. 4: 307–310.

Das, P. (1996). Conservation of Himalayan Mahseer. *Fish. Chimes.*, **15(10)**: 13–15.

Das, P. and W.S. Lakra (1992). Conservation of fish genetic diversity in India. In: Khoshoo, T.N. and Manju Sharma (eds.) *Sustainable Management of Natural Resources*, Malhotra Pub. House. New Delhi, pp. 285–299.

Das, S.M. (1959). The scales of fresh water fishes of India and their importance in age determination and systematic. *Proc. Inst. All India Congr. Zool.* **2**: 621–629.

Das, S.M. (1967). Studies on the gonads and fecundity of food fishes of Kashmir. *Proc. Nat. Acad. Sci.*, **8**: 111–120.

Das, S.M. and Moitra, S.K. (1955). Studies on the food of some common fishes of U.P. India Part-II. On the types of fish food and the variations on the relative length of alimentary canal with description of later. *Proc. Nat. Acad. Sci.*, **26(4)**: 213–223.

Das, S.M. and Moitra, S.K. (1958). On the feeding types of fishes and the variations in the alimentary canal in relation to food. *Jour. Ichthyo.*, (Moscow), **10**: 29–40.

Das, S.M. and Moitra, S.K. (1956). The surface the mid and the bottom feeding fishes of U.P. *Proc. India. Sci. Congr.*, **3**: 307.

Das, S.M. and Moitra, S.K. (1965). Studies on food of twenty four species of fishes of U.P. *Ichthylogica.*, **4(2)**: 107–116.

Das, S.M. and Pathani, S.S. (1978). Studied on the biology of Kumaon mahseer, *Tor putitora. Indian J. Anim. Sci.*, **48(6)**: 461–465.

Das, S.M. and Srivastava, V.K. (1959). *Proc. Nat. Acad. Sci. India,* **29**: 174.

Das, S.M., S.K. Mishra and S.K. Moitra (1955). Feeding habit of fresh water fishes of Uttar Pradesh. *Curr. Sci.*, **24**: 4.

Dasgupta, M. (1982). *An investigation on some aspects of the biology of mahseer from the North-Eastern India.* Ph.D. Thesis North-Eastern Hill University, Shillong.

Dasgupta, M. (1986) A study on the food and feeding habits of the copper mahseer *Acrossocheilus hexagonolepis* (McClelland). *Indian J. Fish.*, **35**: 92–98.

Dasgupta, M. (1987). A study on the maturity and spawning of *Acrossochelius hexagonolepis* (McClleland) from North-Eastern India. *Proc. 74th Ind. Sc. Cong.*, **3**: 123.

Dasgupta, M. (1989a). Biometry of the copper mahseer *Acrossocheilus hexagonolepis* (McClleland) from the North-Eastern India. *Avg. Mus. Roc.*, **1(25)**: 361–374.

Dasgupta, M. (1989b). Biometry of the mahseer *Tor tor* (Hamilton). *Proc. 76th Ind. Sc. Cong.*, **3**: 8.

Dasgupta, M. (1990a). A study on the food and feeding habits of the mahseer *Tor tor* (Hamilton). *Indian J. Fish.*, **37(4)**: 297–304.

Dasgupta, M. (1990 b). A study on the food and feeding habits of the mahseer *Tor putitora* (Hamilton). *Proc. 80th Ind. Soc. Cong.*, **3**: 22–23.

Dasgupta, M. (1991a). Length-weight relationship and condition factor of mahseer, *Tor putitora* (Hamilton) from Garo Hills, Meghalaya. *Indian J. Fish.*, **38(1)**: 35–38.

Dasgupta, M. (1991b). Biometry of the mahseer *Tor putitora* (Hamilton) collected from Garo Hills, Meghalaya, India. *Indian J. Fish.*, **38(2)**: 129–131.

David, A. (1953). On some new records of fish from the Damodars and the Mahanadi river system. *J. Zool. Soc. Ind.*, **5(2)**: 12–19.

David, A. (1955). Notes on bionomics and some early stages of the Mahanadi mahseer. *J. As. Soc. Bengal*, **19(2)**: 197–289.

David, A. (1959). Observations on some spawning grounds of the Gangetic major carps with note on carp seed resources in India. *Indian J. Fish.*, **6**: 327–341.

David, A. (1963a). Sexual dimorphism, fecundity and food of the estuarine bagrid *Mystus gulio* (Ham.). *Proc. Nat. Acad. Sci. India*, **331(3)**: 385–410.

David, A. (1963b). Fishery and biology of the Schilbeid catfish, *Pungasius pungasius* (Hamilton) its utility and propagation in culture ponds. *Indian J. Fish.*, **10(2)**: 521–600.

David, A.N., G.S. Rao and M.F. Rehman (1970). A note on the herbivorous feeding habit of *Puntis pulchellus* (Day). *J. Ins. Fish. Soc. India.*, **2**: 159–160.

Day, F. (1871). Monograph of Indian Cyprinidae (Part II). *J. As. Soc. Bengal*, **11(2):** 77.

Day, F.(1873). *Report on freshwater fish and fisheries of India and Burma.* Calcutta Government Printing Press, p. 118.

Day, F. (1876–1879). *The fishes of India, Burma and Ceylon being a natural history of fishes known to in habit sea and fresh water.* William Davson, London. p. 788.

Day, F. (1878). *The Fishes of India.* William Dawsan and Sons Ltd., pp. 590–591.

De, D.K., M. Sinha and A. Ghosh (1994). Impact of Farakka barrage on the spawning of *Hilsatenualosa ilisha* (Hamilton) in the Hoogly eastuary. *J. Inland Fish. Soc. India*, **26(1)**: 121–124.

Desai, V.R. (1970). Studies on fishery and biology of *Tor tor* (Hamilton) from river Narmada. I. Food and feeding habits. *J. Inland Fish Soc. India*, **2**: 101–112.

Desai, V.R. (1973). Studies on fishery and biology of *Tor tor* (Ham.) from river Narmada. *Proc. Indian Nat. Sci. Acad.*, **39**: 228-248.

Desai, V.R. (1992a). Fishery of Sukta reservoir (Distt. Khandwa: Madhya Pradesh) in the context of mahseer fish. *Punjab Fish Bull.*, **16(1)**: 37–43.

Desai, V.R. (1992b). Food of *Tor tor* (Ham.) from four localities of India. A comparative analysis. *Punjab Fish. Bull.*, **16(1)**: 45–48.

Desai, V.R. (1993). Reproduction biology and spawning ecology of *Tor tor* (Ham.) from river Narmada. In: *Coldwater Fisheries of India*, Garhwal University, Srinagar–Garhwal (U.P.).

Desai, V.R. and S.J. Karamchandani (1968). A note on the food and feeding habits of *Tor* mahseer, *Tor tor* (Ham.) from river Narmada. *J. Bombay Nat. Hist. Soc.*, **65(2)**: 493-495.

Dhanze, J.R. and R. Dhanze (1996). Impact of habitat shrinkage on the indigenous fish genetic resource of Beas drainage system, (H.P.). *Proceeding symposium on fish genetic and biodiversity conservation for sustainable production*. NBFGR, 9-10.

Dobriyal, A.K. and Bahuguna, A.K. (1987). Morphometric characters and their relationship in the hillstream loach, *Loe Macheilus montanus*. *Him. J. Env. Zool.*, **1**: 23-27.

Dobriyal, A.K. and Singh, H.R. (1989). Spawning ecology of hill stream minor carps *Puntius chilinodes*. *Ad. Bios.*, **8**: 139–148.

Dobriyal, A.K. and Singh, H.R. (1998). Ecological basis for ichthyofaunal variation in two hillstream of Garhwal Himalaya. In: Joseph, M. (Ed.) *The First Indian Fisheries Forum Proceeding*. Asian Fisheries Society, Indian Branch, pp. 313–317.

Dubey, G.P. (1986). Mahseer reservoir of Madhya Pradesh and its conservation with special reference to Narmada basin. *Pbunjab Fish. Bull.*, **10**: 12–18.

Dubey, Kiran and G.P. Dubey (1986). Biometric studies of Indian mahseer, *Tor tor* (Ham.) from Narmada river. *Matsya*, **12-13**: 126–132.

Dunsford, H.S. (1911). *The report on the fish and the fisheries of the Punjab.* Government Press, Lahore, p. 23.

Dwivedi, S.N. and M.R. Menezes (1974). A note on the morphometric and ecology of *Brochirus orientalis* (Block and Schneider) in the estuary of Goa. *Geobios.*, **1(4)**: 80–83.

Edmondson, W. (1959). *Rotifera fresh water biology* 2nd edn. Wiley and Sons, New York, pp. 420–484.

Edmondson, W.T. (1974). A Simplified method for counting phytoplankton. In: *"A manual on methods for measuring primary production in aquatic environment"* Vollen Weider, R.A. (Ed.) Oxford, pp. 14-16. Oxford, Blackwell Sci. Publication, p. 1387.

Ellis, M.M. (1937). Detection and measurement of stream pollution. *Bull. Vs. Fish.*, **48**: 365–437.

Erickson, C.M. (1983). Age determination of Manitoban walleyes using otoliths, dorsal spines and scales, North America. *J. Fish. Manage.*, **3**: 176–181.

Frost, W.E. (1997). The food of charr, *Salvelinus nellighbi* (Gunther) in Wintermere. *J. Fish. Biol.* **11**: 531–547.

Frost W.E. and Kipling, C. (1959). The determination of the age and growth of pike (*Esox lucis* L.) from scale and opercular bones. *J. Cons. Perm. Int. Explor. Mer.*, **24**: 314–341.

Ganapati, S.V. and Alikunhi, K.H. (1950). Notes on the spawning of carps in river Tungabhadra in response to the off-screen freshts. *J. Zool. Soc. India*, **2**: 93–95.

Garham, M.(1929). Studies of age determination in fish Pt. II. *Fish. Invest. London*, pp. 1–50.

Gaur, S.K. and Pathani, S.S. (1996). Studies on the spawning ecology of a hillstream carp, *Barillius vagra* (Hamilton) from Kumaon Himalaya. *Him. J. Env. Zool.*, **10**: 63–66.

George, K.C. (1959). A method for distinguished the sex of the oil sardine, *Sardinella longiceps* in the field. *Indian J. Fish.*, **6(2)**: 322–326.

George, K.C. (1963). Selective digestion in the major carps of India. *Curr. Sci.*, **32**: 77-80.

George, K.C. (1965). The food of the major carps in some fish ponds in Delhi. *India J. Fish.*, **12(1)**: 25–59.

Gjosaeter, J.; Dayaratne, P.; Bergstad, D.A.; Gjosaeter, H.; Sousa, M.I.; and Beck, I.M. (1984). Ageing of tropical fish by growth rings in the otoliths. *FAO Fish. Circ.*, **776**.

Gould, D.S.J. (1966). Allometry and size in outogeny and phylogeny. *Society Biological Review of Cambridge Philosophical.* **41**: 587–640.

Gunther, S.D. and Rer Nat. Habil (1959). *Fresh water fishes of the world, Vols. I and II.* Translated by Tucker, D.W. (1973), T.F.M.N. Publication Inc. Ltd. London, pp. 645.

Gupta, Poonam and Khan, H.A. (1992). Mahseer fisheries in India. In: National Seminar on *Endagered Fishes of India,* held at National Bureau of Fish Genetic Resources, Allahabad. (April 25-26, 1992, Abstract no. 64), pp. 79–80.

Gupta, P. and H.A. Khan (1994). Mahseer fisheries in India. In.: P.V. Deharai, P. Das and S.R. Verma (eds.). *Threatened Fishes of India,* Natcon Pub. No. 4, pp. 181–190.

Hamilton, B. (1822). *An account of the fishes found in the river Ganges and its Branches,* Bishan Singh M.P. Singh Publication Dehradun, p. 541.

Hannook, R.S., J.W. Jones and R. Shaw (1976). A preliminary report on the spawning behaviour and nature of sexual selection in the barbel, *Barbus barbus* (L.). *J. Fish. Biol.*, **9**: 121–128.

Hartley, A.H. (1947). Food and feeding relationship in community of fresh water fishes. *J. Anim. Ecol.*, **33**: 1–14.

Hickling, C.F. and Rutenberg, E. (1936). The ovary as an indicator of spawning period of fishes. *J. Mar. Ass. U.K.*, **21**: 311–317.

Hile, R. (1970). Body scale relation and calculation of growth in fishes. *Trans. Am. Fish. Soc.*, pp. 468–474.

Hora, S.L. (1921). On some new or rare species of fishes from the Eastern Himalayas. *Rec. India Mus.*, **22(5)**: 731–744.

Hora, S.L. (1939). The Game Fishes of India VIII. The mahseer or the large scale barbels of India. I. The *putitor* mahseer, Barbus (*Tor*) *putitora* (Hamilton). *J. Bombay Nat. Hist. Soc.*, **41**: 272–285.

Hora, S.L. (1940a). Dams and the problem of migratory fishes. *Curr. Sci.*, **9**: 406–407.

Hora, S.L. (1940b). The Game Fishes of India IX. The mahseer or the large scaled barbels of India. 2. The Tor mahseer Barbus (*Tor*) *tor* (Hamilton). *J. Bombay Nat. Hist. Soc.*, **41**: 518–525.

Hora, S.L. (1940c). The game fishes of India, X. The mahseers or the large scaled barbels of India. 3. The Mosal mahseer Barbus (*Tor*) *mosal* (Hamilton). *J. Bombay Nat. Hist. Soc.*, **41**: 784–794.

Hora, S.L. (1942a). The effect of dams on the migration of the hilsa fish in Indian water. *Curr. Sci.*, **11**: 470–471.

Hora, S.L. (1942b). The Game Fishes of India. XV. The mahseers or the large scaled barbels of India. 8: On the specific identity of Sykes species of Barbus from Deccan. *J. Bombay Nat. Hist. Soc.*, **43(2)**: 163–169.

Hora, S.L. (1943). The game fishes of India, XVI. The mahseers or the large scaled barbels of India. 9. Further observations on mahseers from the Deccan. *J. Bombay Nat. Hist. Soc.*, **44**: 1–8.

Hora, S.L. (1945). Analysis of factors influencing the spawning carps symposium of the *"Factors influencing the spawning of Indian carps." Proc. Nat. Inst. Sci. India*, **11**: 301–311.

Hora, S.L. (1951). Knowledge of ancient Hindus concerning fish and fisheries of India. 2. Fish in the Sutras and Smriti Literature. *Jour. Asiatic Soc. Letters*, **17**: 61–68.

Hora, S.L. (1953). Knowledge of the Ancient Hindus concerning fish and fisheries of India. 4. Fish in the Sutras and Smriti Literature. *Jour. Asiatic Soc. Letters*, **18**: 63–67.

Hora, S.L. and Mukerjee, D.D. (1936). Fish of the Eastern Doons united provinces with introduction and remarks on mahseer fisheries. *Rec. Indian Mus.*, **38**: 133–146.

Hora, W.S. (1957). *"Physiology of Fishes"*. Academic Press Inc., New York, p. 640.

Hubbs, C.L. (1921). Geographic variation of *Notemigenus cryoluscar* an American minnow. *Trans. Illinois St. Acad Sci.* **11**: 147–151.

Hussain, A. (1945). Analysis of experimental data on the spawning of carps in the Punjab. *Proc. Inst. Sci. India*, **11**: 320–327.

Hyatt, H.D. (1979). Feeding strategy. In: *Fish Physiology, Vol 8: Bioenergetics and Growth*, W.S. Hora; D.J. Randwall and J.R. Brett, Eds., Academic Press, New York and London, p. 765.

Hynes, H.B.N. (1950). The food of freshwater sticklebacks (*Gasterosteus aculeatus* and *Pygotius pungistius*) with a review of the methods used in the studies of food of fishes. *J. Anim. Ecol.*, **19(1)**: 35–58.

Iftekhar, M. and S.N. Dwivedi (1977). A note on the morphometry of *Pseudorhombus triocellatus* (Gunther). *Bulletin of the Central Institute of Fisheries Education*, pp. 195–198.

Ishiwata, A. (1969a). Ecological studies on the feeding of fishes VIII. Frequency of feeding and satiation amount. *Bull. Jpn. Soc. Sci. Fish.*, **35**: 979–984.

Ishiwata, A. (1969b). Ecological studies on the feeding of fishes: VIII. Frequency of feeding and growth. *Bull. Jpn. Soc. Sci. Fish*, **35**: 985–990.

Ivlev, V.S. (1961). *"Experimental Ecology of the Feeding of Fishes"*. (Transl. D. Scott) Yala University Press, New Haven, p. 710.

James, P.S.B.R. and Baragi, V.M. (1980). Ovary as an indicator of frequency of spawning in fishes. *Proc. Indian Nat. Sci. Acad.*, **846**: 149–159.

Jan, N.A. and S.M. Das (1970). Quantitative and qualitative studies on the food of eight fishes of Kashmir valley. *Ichthyologica*, **10(1–2)**: 20–26.

Jayaram, K.C. (1981). *The fresh water fishes of India, Pakistan, Bangladesh, Burma and Sri Lanka*. A handbook, Zoological Survey of India, Calcutta, pp. 730.

Jayaram, K.C. (1987). *The fresh water fishes of India, Pakistan, Bangladesh, Burma and Sri Lanka*. Zoological Survey of India, Calcutta, pp. 480.

Jhingran, A.G. (1971). Validity of scales as age indicator in *Septipinna phasa* (Ham.). *Indian J. Fish.*, **8**: 291–311.

Jhingran, A.G. (1972a). Diet composition of *Gudusia chapra* (Ham.) from Ganga river system. *Proc. Indian Nat. Sci. Acad.*, **38B(1–2)**: 94–105.

Jhingaran, A.G. (1972b). Fluctuations in the ponderal index of the Gangetic anchoy *Septipinna phasa* (Ham.). *J. Inl. Fish. Soc. India*, **4**: 1–9.

Jhingran, A.G. (1990). *Restoration of fisheries in the Ganga river system*. Working paper for the meeting of the technical sub committee of the co-ordination committee on restoration of fisheries in Ganga river system, p. 19.

Jhingran, V.G. (1957). Age determination of the Indian major carp, *Cirrhinus mrigala* (Ham.) by means of scales. *Nature*, **179**: 468–469.

Jhingran, V.G. (1968). Synopsis of biological data on *Catla catla* (Hamilton). *F.A.O. Fisheries synopsis*, **32**: 1-30.

Jhingran, V.G. (1975). *Fish and Fisheries of India.* Hindustan Publishing Corporations, New Delhi, India, p. 754.

Jhingran, V.G.(1982). *Fish and Fisheries of India.* Hindustan Publishing Corporations, New Delhi, India, pp. 931.

Jhingran, V.G. (1991). *Fish and Fisheries of India*, 3rd Edition. Hindustan Publishing Corporation, New Delhi, India, p. 985.

Jhingran, V.G. and Sehgal, K.L. (1978). *Coldwater Fisheries of India.* Inland fisheries Society of India, Central Inland Fisheries Research Institute, Barrackpore, p. 239.

Johal, M.S. and Tondon, K.K. (1981a). Age and growth of *Cirrhina mrigala* (Pisces: Cyprinidae) from northern India. *Vest Sc. Spolec Zool.*, **51**: 252–280.

Johal, M.S. and Tondon, K.K. (1981b). Age, growth and length-weight, relationship of *Tor putitora* from Govindsagar, Himachal Pradesh, India. *Pub. Fish. Speci. Publ.* Coldwater Fisheries Issue, pp. 43–48.

Johal, M.S. and Tondon, K.K. (1983). Age, growth and length-weight relationship of *Catla catla* and *Cirrhina mrigala* (pisces) from Sukhna lake, Chandigarh (India). *Vest. Cs. Spolac. Zool.*, **47**: 87–98.

Johal, M.S. and K.K. Tondon (1985). Use of growth parameters in *Labeo rohita* (Pisces: Cyprinidae). *Vest. Cs. Spolac, Zool.*, **49**: 101–107.

Johal, M.S. and Tondon, K.K. (1992). Age and growth of the carp *Catla catla* (Hamilton, 1822) from northern India. *Fish. Res.*, **14**: 83–90.

Johal, M.S., K.K. Tondon and G.S. Sandhu (1999). Age and growth of an endangered cold water fish–golden mahseer, *Tor putitora* (Hamilton) from Govindsagar, Himachal Pradesh, India, pp. 59–73: *Ichthyology: Recent research advances* (D.N. Saksena ed.). Oxford & IBM Publishing Co. Pvt. Ltd., New Delhi.

Jones, S. and A.G.K. Menon (1958). Spawning of *Hilsa ilisa* (Hamilton) in the Hoogly. *Science and Culture*, **15(11)**: 443–444.

Joshi, B.D., R.C.S. Bisht, Namita Joshi and Rajendra Singh (1996). A study of plankton and benthic components of three selected tributaries of river Ganga between Devprayag and Rishikesh. *Him. J. Env. Zool.*, **10**: 23–26.

Joshi, C.B. (1980). Procurement of stocking material of mahseer, *Tor putitora* from the hill stream in U.P. and Himachal Pradesh with special reference to collection technique. *Proc. IV AISI Lot, J. Acad. Ichytyho.*, **15**: 169–172.

Joshi, C.B. (1984). A special method of fishing for juvenile of mahseer, (*Tor putitora*) in the stream of Himachal Pradesh. *J. Bombay Nat. Hist. Soc.*, **80(2)**: 426–427.

Joshi, C.B. (1988). Induced breeding of mahseer *Tor putitora* (Hamilton). *J. Ind. Fish. Soc.*, **20(1)**: 66–67.

Joshi, C.B. and Kuldeep Kumar (1980). Food spectrum of juveniles of Golden mahseer (*Tor putitora*) collected from the streams of Jammu province. *J. Inland Fish, Soc. India*, **12(2)**: 105–108.

Joshi, C.B. and Shyam Sunder (1995). Breeding and culture of snow trout *Schizothorox richardsonii* (Gray) in Kumaon Himalayas. *Uttar Pradesh J. Zool.*, **15(2)**: 136–140.

Joshi, C.B.; K.L. Sehgal and Shyam Sunder (1978). Observation on the fishery resources of the hillstreams of Jammu province with special reference to mahseer and other commercially important species. *Indian J. Fish.*, **25**: 197–206.

Joshi, K.D.; Mahanta, P.C., Srivastava, S.M. and Patiyal, R.S. (1994). *Problem of mass mortality of fishes of Agriculture Imp. and their management*, pp. 19–21.

Joshi, P. (1994). Status of fish conservation in river Ramganga. In: Dehadrai, P.V.; P. Das and S.R. Verma (eds.) *Threatened Fishes of India*, Natcon Pub. No. 4: 149–153.

Joshi, S.N.; G. Tripathi and H.C. Tiwari (1993). Fish and fisheries of the Gori Ganga In: *Advance in Limnology* (ed. H.R. Singh), Narendra Publishing House Delhi, pp. 361–366.

Kagwade, P.V. (1971). Age and growth of *Polynemus haptadactylus* (Cuv. and Val.) from the Bombay, Saurashtra region. *Indian J. Fish.*, **1(1-2)**: 33–51.

Kapoor, B.G.; Smith, H. and Verghina, J.A. (1975). The alimentary canal and digestion in teleosts. *Adv. Mar. Biol.* **13**: 109–139.

Kapoor, D.; P.C. Mahanta and A.K. Pandey (1998). Ichthyodiversity of India, Status & Conservation. In: A.G. Ponniah; P. Das & S.R. Verma (eds.) *Fish Genetics Biodiversity and Conservation*, Natcon Pub. No. 5: 47–50.

Karamchandani, S.J. and M.D. Pisholkar (1967). On the occurrence and breeding of *Catla catla* (Hamilton) in Tapti river. *J. Bombay Nat. Hist. Soc.*, **64(2)**: 375–377.

Karamchandani, S.J. and Pisholkar, M.D. (1967). Survey of the fish and the fisheries of the Tapti river, Survey Rep. *Cent. Inl. Fish Res. Inst.*, Barrackpore, **4**: 29.

Karamchandani, S.J.; V.R. Desai; M.D. Pisholkar and G.K. Bhatnagar (1967). Biological investigation on the fish and fisheries of Narmada river (1958–66) Bull Centre. *Inl. Fish. Res. Inst.* Barrackpore, **10**: 40.

Khan, H. (1939). Study of the sex organ of mahseer. *J. Bombay Nat. Hist. Soc.*, 41(1): 231–243.

Khan, Hamid (1940). Fish ladders in Punjab. *J. Bombay Nat. Hist. Soc.*, **41(3)**: 551–562.

Khan H. (1942). Spawning of carps and their spawning grounds in the Punjab. *J. Bombay Nat. Hist. Soc.*, **43**: 416–427.

Khan, H. (1945). Observation on the spawning behaviour of carps in Punjab. In the symposium on the factors influence the spawning of Indian carps. *Proc. Nat. Inst. Sci. India*, **11**: 324–327.

Khanna, D.R. and Badola, S.P. (1990). Ichthyofauna of the river Ganga at the foothills of Garhwal Himalaya. *J. Nat. and Phy. Sci.*, **4(1–2)**: 153–162.

Khanna, D.R.; D.S. Malik and D.S. Rana (1998). Phytoplanktonic communities in relation to certain physico-chemical parameters of Ganga canal at Hardwar. *Him. J. Env. Zool.*, **12**: 193–197.

Khanna, D.R.; D.S. Malik and Rupendra (1998). Fish fauna of the river Ganga at Rishikesh (U.P.). *J. Natcon.* **10(2)**: 211–213.

Khanna, D.R.; D.S. Malik and S.P. Badola (1997). Population of green algae in relation to physico-chemical factors of the river Ganga at Lal-Ji-Wala, Hardwar, Uttar Pradesh. *Indian J. Ecol.*, **17(8)**: 237–240.

Khanna, D.R.; S.P. Badola and Anoop K. Dobriyal (1993). Plankton ecology of the river Ganga at Chandighat, Hardwar. In: *Advances in Limnology* (H.R. Singh ed.), pp. 171-174.

Khanna, D.R.; Seth, T.R.; Malik, D.S. and Singh R.K. (1999). Fishes and their ecology of the river Suswa at Raiwala, Dehradun. *Proceedings of National Conference in Sustainable Eco. Sys and Env.*, pp. 105–110.

Khanna, H.A. and V.G. Jhingran (1975). Synopsis of biological data on Rohu, *Labeo rohita* (Hamilton). *F.A.O. Fisheries Synopsis*, **111**: 1–100.

Khanna, S.S. (1961). Alimentary canal in some Teleosten fishes. *J. Zool. Soc. India*, **13(2)**: 206–219.

Khanna, S.S. and M.C. Pant (1964). Digestive tract and the feeding habits of some teleosten fishes. *Agra University J. Res. Sci.*, **13(1)**: 15–30.

Khillare Yashvant and Krzystof Skora (1998). Observation on otolith and age in eel-pout, *Zoarces viviparus*, from Hel Peninsula, Poland in relation to reproduction. *J. Indian Fish.*, **10(2)**: 243–251.

Kirka, A. (1974). Variation in somatic characters of the brown trout (*Salmo trutt fario* Linn., 1758) in the headwaters of the Poprad river (High Tatra Mts.). *Zoological List*, **23(1)**: 85–94.

Kohlar, A.C. (1960). The growth, length-weight relationship and maturity of haddock (*Malanogammus aeglefinus* L.) from the refion of Lockeport N.S. *J. Fish Res. Bd., Canada*, **17(1)**: 41–60.

Krishnamurthy, K.N., B.V. Govind and G.K. Bhatnagar (1964) Experimental fishing with gil nets in the Tungabhadra reservoir. *Indian J. Fish.*, **11(1)**: 465–478.

Kulkarni, C.V. (1971). Spawning habits, eggs and early development of Deccan mahseer, *Tor khudree* (Sykes). *J. Bombay Nat. Hist. Soc.*, **67(3)**: 510–521.

Kulkarni, C.V. (1979). Air transport of mahseer (Pisces) eggs in moist cotton wool. *Aquaculture*, **16**: 367–368.

Kulkarni, C.V. (1986). Methods of conservation of mahseer resources. *Pb. Fish. Bull.*, **10(2)**: 33–35.

Kulkarni, C.V. (1992). Role of aquaculture in the conservation of mahseer fisheries. In K.L. Sehgal (ed.). *Recent Researches in Coldwater Fisheries*, pp. 173–176, Today and Tomorrows Printers & Publishers, New Delhi, India.

Kulkarni, C.V. and Ogale, S.N. (1991). *Conservation of the mighty mahseer of India.* Thrid Workshop on conservation and artificial propagation of mahseer. Tata Electrical Companies Lonawala, Maharashtra, August (1991), p. 35.

Kumar, Kuldeep (1986). Ecological studies of mahseer streams of Assam with special reference to the development of sport fisheries. *Pb. Fish. Bull.*, **10**: 37–44.

Lackey, J.B. (1938). U.S. Public Health Reports, **53**: 2080–2093.

Lagler, K.F. (1959). Freshwater Fishery Biology. W.M.C. Brown Co., Lowa. Ed. 2, p. 421.

Lake, J.S. (1957). Trout populations and habitat in New South waters. *Aust. J. Mar. Freshwater Res.*, **8**: 414–450.

Lal, M.B. (1962). Adaptation of the Indian hill stream fish *Tor putitora*. *DAV College Dehradun Magazine*, **29(1)**: 49–50.

Lal, M.B. (1963). Distribution and migration of *Tor putitora* (Ham.). *Cheetal*, **1(6)**: 54–56.

Lal, M.B. (1968). Studies on the anatomy and histology of the alimentary canal of a carp *Tor putitora* (Ham.). *Proc. Nat. Acad. Sci. India*, **38B(11)**: 127–136.

Lal, M.B. and Chaterjee, P. (1962). Survey of Eastern Doon fishes with certain notes on their biology. *J. Zool. Soc. India*, **14(1–2)**: 230–243.

Langer, R.K. (1987). Food and feeding habits of *Schizothorax longipinnis* (Heckel) from Jhelum river, Kashmir. *J. Ind. Fish. Assoc.*, **14&15**: 41–48.

Langer, R.K.; Somalingam and U.K. Maheshwari (1992). Availability of mahseer (*Tor tor*) seed in river Narmada at Hoshangabad

(M.P.) and its feeding habits and growth in captivity (farm Pond, Powarkheda). In: *Recent Researches in Coldwater Fisheries* (K.L. Sehgal ed.), Today and Tomorrows Printers and Publishers, New Delhi India. pp. 145–156.

Lear, W.H. (1969). Length-weight relationship of commercial size Greenland halibul *Reinhardtius hippoglossoides* (Walbaum) Internation comm. *North-west Atlantic Fish. Res. Bull.*, **6**: 191–121.

Lear, W.H. and T.K. Pitt (1975). Otolith age validation of Greenland halibul (*Reinhardtius hippoglossoides*). *J. Fish. Res. Bd. Canada*, **32(2)**: 289–292.

Le Cren, E.D. (1951). The length weight relationship and seasonal cycle in gonad weight and condition in the perca (*Perca fluviatilis*). *J. Anim. Ecol.*, **20**: 201–219.

Lee Rosa, M. (1920). A review of the methods of age and growth determination in fishery by means of scales. Min. Agric. and Fish. *Fish Invest Ser-II* **4(2)**: 1–32.

Libosvarky, J. and Ruban, G.I. (1985). In search of morphometric deviations in Roach from heated waters. *Folia Zoologica.*, **34(3)**: 267–278.

Liley, N.R. (1982). Chemical communication in fish. Can. *J. Fish. Aquat. Sci.*, **39**: 22–35.

Lindsey, G.C. (1988). Controlling meristic variatons. In: Hoar and D.J. Randall, *Fish Physiology* Vol. XB, Academic Press Inc., pp. 197–274.

Lowe McConnel, R.H. (1971). *Identification of fresh water fishes methods for assessment of fish production in freshwaters*, pp. 45-81. (Ed.) Ricker, W.E. Blackwell Scientific Publication, Oxford and Edinburg.

Mahajan, C.L. (1961). Fish fauna of Muzaffarnagar district, Uttar Pradesh. *J. Bombay Nat. Hist. Soc.*, **62(3)**: 440–454.

Maheshwari, U.K.; Ibodi Singh; Bhabtosh Ray; S.K. Bhatnagar and Sushma Nigam (1987). Observation on food and feeding habits of silver carp, *H. molitrix* (C&V). *MSS*, pp. 25–29.

Malik, D.S. (1994). Feeding behavior of fingerlings of mahseer, *Tor putitora* (Ham.). *Biosphere*, **6(1)**: 25–31.

Malik, D.S.; K.S. Negi and M. Bhanot (2002). Spawning periodicity of mahseer *Tor putitora* in relation to water quality in river Ganga. In: Das, P.; J.R. Dhanze; S.R. Verma and D.S. Malik (eds.). *Coldwater Fish Genetic Resources and their Conservation,* Natcon Pub No. 7: 253–260.

Malik, D.S.; K.L. Sehgal and D.R. Khanna (1998). Ichthyodiversity and conservation of endangered fish species in upland waters. *Him. J. Env. Zool.* **12**: 199–204.

Mann, R.H.K. (1976). Observation on the age, growth, reproduction and food of the chub, *Cqualis cephalus* (L.) in river Stour dorset. *J. Fish. Biol.,* **8(3)**: 265–288.

Marr, A.C. (1955). The use of morphometric data in systematic and relative growth studies in fishes. *Copeia.,* pp. 23-30.

Martin, N.V. (1954). Catch and winter food of lake trout in certain *Alongwin park* Lakes. *J Fish Res. Bul. Ottawa* (Canada), **11**: 5–10.

Martin, R.L. (1966). A dissection of the weberian canal. *Turtox News,* **39(8)**: 178–182.

Mathur, P.K. (1964). Studies on the maturity and fecundity of the *Hilsa ilisa* (Ham.) in the upper stretches of Ganga. *Indian J. Fish.,* **11(1)**: 423–448.

McClleland, J.(1839). Indian Cyprinidae. *Asiat Researches,* **19**: 217–468.

McClleland, J. (1844). Description of a collection of fishes made at Chusan and Ningpo in China. *Calcutta Jour. Nat. Hist. Soc.,* **4**: 399.

Mc Donald, A.S.J. (1936). A Fishing trip in Kumaon. *J. Bombay Nat. Hist. Soc.,* 38(3): 598–600.

Mc Donald, A.S.J. (1948). Circumventing the mahseer and other sporting fishes in India and Burma. *J. Bombay Nat. Hist. Soc.,* Bombay, 306.

McHugh J.L. (1954). Geographic variations in pacific herring. *Copeia.,* **2**: 139–151.

Menon, A.G.K. (1962). A distribution list of fishes of the Himalayas *J. Zool. Soc. India.* **14(1–2)**: 23–32.

Menon, A.G.K. (1988). Conservation of Ichthyofauna of India. In: Arun, G. Jhingran and V.V. Sugunan, Ed., *Conservation and Management of Inland Capture Fishery Resources*, pp. 20–24.

Menon, A.G.K. (1991). Taxonomy of mahseer fishes of the genus *Tor* Gray with description of a new species from the Deccan. *J. Bombay Nat. Hist. Soc.*, **89(2)**: 210–228.

Menon, M.D. and P.I. Chako (1958). The food and feeding habits of some fresh waters fishes of Madras state. *J. Bombay Nat. Hist. Soc.*, **35(1)**: 117–124.

Mishra, K.S. (1950). Sexual dimorphism in *Labeo buggut* (Sykes). *J. Zool. Soc. India*, **2(3)**: 91–92.

Mishra, K.S. (1959). An aid to the identification of common fishes of India and Pakistan. *Rec. Indian Mus.*, **57(1–4)**: 1–320.

Mishra, K.S. (1962). An aid to identification of the common commercial fishes of India and Pakistan. *Rec. Indian. Mus.*, **57(1–4)**: 119.

Mishra, K.S. (1967). The fauna of India and the adjacent countries, 2nd edn. *Pieces*, **3(21)**: 367.

Mishra, R.N. (1953). On the gut contents of *Labeo rohita* (Ham.), *Cirhina mrigila* (Ham.), *Catla catla* (Ham.). 42nd *Proc. Ind. Sci. Cong.*, **3**: 210.

Mitra, P.M.; G.C. Laha and Ghosh, K.K. (1977). Projections of expected fish catches from Hoogly-Matlah eusterine system during 1976-80. *J. Inland Fish Soc. India*, **9**: 131–137.

Mohan Devendra and Punit Saraswat (2000). Length-weight relationship of *Cyprinus carpio* from Rawti Pond, Jodhpur. *J. Natcon*, **12(2)**: 211–216.

Moriarty, C. (1973). A technique for examining eel otolith. *J. Fish Biol.*, **5(2)**: 181–183.

Mraz (1964). Age, growth, sex ratio and maturity of the witefish in central green bay and adjacent waters of lake Michigan. *Fishery Bull.*, **63(3)**: 619–634.

Mukherjee, H.K. and S.N. Ghosh (1945). Food of major carps. *Proc. Indian Sci. Congr.*, **32(3)**: 110–111.

Mukherjee, H.K. and Sengupta, H.N. (1946). Correlation between food, body weight and length of gut in *Cirrhina reba. Proc. Ind. Sci. Cong.*, **33(3)**: 16.

Nar Simha, K.A. (1970) On the length-weight relationship and relative condition in Trichirus L. *Indian J. Fish.*, **17(1–2)**: 90–96.

Nath, M.M.; R.K. Jasbir Singh and M.M. Chopra (1994). The report on the decline of sport fish in the rivers and streams of the hills of Uttar Pradesh with particular reference of the Doon valley region. In: Dehadrai, P.V.; P. Das and S.R. Verma (eds.). Natcon Pub No. 4: 219–227.

Nautiyal, P. (1983). Some aspects of bioecology of *Tor putitora* (Hamilton) in relation to hydrobiology of some hillstream. Ph.D. Thesis, Garhwal University Srinagar (Garhwal).

Nautiyal, P. (1984). Natural history of the Garhwal Himalayan mahseer *Tor putitora* (Hamilton). II. Breeding biology. *Proc. Ind. Acad. Sci.*, **93**: 97-106.

Nautiyal, P. (1985). Length-weight relationship and relative condition factor of the Garhwal Himalayan mahseer with reference to its fishery. *Ind. J. Ani. Sci.*, **55**: 65–70.

Nautiyal, P. (1989). Mahseer conservation–Problems and prospects. *J. Bombay Nat. Hist. Soc.*, **85**: 32–36.

Nautiyal, P. (1990). Natural history of Garhwal Himalayan mahseer, growth rate and age composition in relation to feeding and breeding ecology. In: R. Hirano and Hanya (eds.). *Proc. 2nd Asian Fishery Forum* (Manila: Asian Fishery Society), pp. 769–772.

Nautiyal, P. and H.R. Singh (1989). River projects and the endangered Himalayan mahseer, *Tor putitora* in the uplands of Garhwal region. A critical evaluation. *Him. J. Env. Zool.*, **3**: 129–134.

Nautiyal, P.; J.P. Bhatt; V.S. Rawat; B. Kishore; R. Nautiyal and H.R. Singh (1988). Himalayan mahseer: Magnitude of commercial fishery in Garhwal hills. In: A.G. Ponniah; P. Das and S.R. Verma (eds.) *Fish Genetic and Biodiversity Conservation*, Natcon Pub. No. **5**: 107–114.

Nautiyal, P. and Lal, M.S. (1982a). Fecundity of Garhwal Himalayan mahseer, *Tor putitora* (Ham.). *J. Bombay Nat. Hist. Soc.*, **82**: 253–257.

Nautiyal, P. and Lal, M.S. (1982b). Recent records of Garhwal Himalayan mahseer, (*Tor putitora*) with a note on its present status. *J. Bombay Nat. Hist. Soc.*, **79**: 639–695.

Nautiyal, P. and Lal, M.S. (1984a). Preliminary observation on the migratory behaviour of the Garhwal Himalayan mahseer. *J. Bombay Nat. Hist. Soc.,* **81**: 204–208.

Nautiyal, P. and Lal, M.S. (1984b). Food and feeding habits of fingerlings and juveniles of mahseer, *Tor putitora* (Hamilton) in Nayar river. *J. Bombay Nat. Hist. Soc.,* **81**: 642–646.

Nautiyal, P. and Lal, M.S. (1985a). Studies on the natural history of Garhwal Himalayan mahseer *Tor putitora* (Hamilton). I maturation. *Indian J. Phys. Nat. Sci.,* **5**: 36–42.

Nautiyal, P. and Lal, M.S. (1985b). Food and feeding habits of Garhwal Himalayan mahseer in relation to certain abiotic factors. *Matsya,* **11**: 31–35.

Nautiyal, P. and Lal. M.S. (1988). Destruction of spawning grounds of mahseer and other fish in Garhwal Himalaya. *J. Bombay Nat. Hist. Soc.,* **85**: 311–314.

NCA (1976). National Commission of Agricultural report. *Ministry of Agriculture,* New Delhi, p. 35.

Needham, G.R. and Needham, P.R. (1966). *A guide to fresh water biology*. Holden Day Inc. San Fransisco, p. 230.

Nevill, C.A. (1915). The breeding habit of the mahseer (Barbus tor). *J. Bombay Nat. Hist. Soc.,* **24(4)**: 838–839.

Nikolosky, G.V. (1963). *"The Ecology of Fishes"* Moscow. (Translation English, Academic Press, London), pp. 1–853.

Om Kumar; Shashi Bisht and Neelu Singh (1990). Studies on water quality vegetation and fish of Song river in eastern Doon valley forest. *Uttar Pradesh J. Zool.,* **10(2)**: 143–147.

Pandey, S.K. and Nautiyal, P. (1997). Statistical evaluation of some meristic and morphometric character of taxonomic significance in *Schizothorax richardsonii* (Gray) and *Schzothorax plagiostomus* (Heckel). *Indian J. Fish.,* **44(1)**: 75–79.

Pathani, S.S. (1977). The problems of Kumaon mahseer, (*Tor tor* and *Tor putitora*). *Uttarakhand Bharti,* **2(1)**: 65–68.

Pathani, S.S. (1978a). A note on secondary sexual characters in Kumaon mahseer, *Tor tor* and *Tor putitora* (Ham.) in Uttar Pradesh. *Indian J. Anim. Sci.,* **48(10)**: 773–775.

Pathani, S.S. (1978b). A note on ecology and conservation of mahseer, *Tor tor* and *Tor putitora* (Ham.) in Kumaon lakes. *Env. India*, **1**: 36–37.

Pathani, S.S. (1979). Otolith as age indicator in the mahseer, *Tor putitora* (Ham.). *Current Science*, **48(21)**: 957–958.

Pathani, S.S. (1980). Studies on the spawning ecology of Kumaon mahseer *Tor tor* (Hamilton) and *Tor putitora* (Hamilton). *J. Bombay Nat. Hist. Soc.*, **79**: 525–530.

Pathani, S.S. (1981a). Age and growth of the mahseer, *Tor putitora* (Ham.) as determines by scales and operculam. *Matsya*, **7**: 41–46.

Pathani, S.S. (1981b). Fecundity of mahseer, *Tor putitora* (Ham.). *Proc. Indian Acad. Sci.*, **90(2)**: 253–260.

Pathani, S.S. (1982). Studies on the fecundity of Himalayan mahseer, *Tor putitora*. *J. Adv. Zool.*, **3(2)**: 122–128.

Pathani, S.S. (1984). Population biology of mahseer *Tor tor* and *Tor putitora* (Ham.). *J. Himalayan Res. Dev.*, **3**: 29–32.

Pathani, S.S. (1988). Recent advances in the biology and ecology of mahseer *Tor tor* and *Tor putitora* (Ham.). In: Nath Surendra; B.D. Sharma and Mrs. Tej Sharma (eds). *Recent Advances in Fish Ecology, Limnology and Eco-conservation*. Creative Publications New Delhi, pp. 115–121.

Pathani, S.S. and L.M. Joshi (1980). On food and feeding habits of fingerlings of the two Kumaon mahseer fishes, *Tor tor* (Hamilton) and *Tor putitora* (Ham.). *Bioresearch.*, **4(1)**: 43–46.

Payer, D.R. and C.G. Scalet (1978). Population and production estimates of fathead min now on South Dakota Prairie wetland. *Frog. Fish Cult.*, **40(2)**: 63–66.

Pillay, T.V.R. (1951). A critique of the methods of study of the food of fishes. *J. Zool. Soc. India*, **4(2)**: 185–196.

Pillay, T.V.R. (1953). Studies on the food and feeding habits and alimentary tract of grey mullet, *Mugil tode* (Forsk.). *Proc. Nat. Inst. Sci. India*, **19(6)**: 77–827.

Pillay, T.V.R. (1957). A morphometric study of the population of Hilsa, *Hilsa ilisha* (Ham.) of the river Hoogly and of Chilka lake. *Indian J. Fish.*, **4**: 344–386.

Pisholkar, M.D. and Karamchandani, S.J. (1981). Fishery and biology of *Tor tor* (Ham.) from Govindgarh lake (Madhya Pradesh) I. Food and feeding habits. *J. Inland Fish Soc. India*, **13(1)**: 15–24.

Pivinica, K. and Hensel, K. (1978). Morphological variation in the genus *Thymalus* Cuvier, 1829 and recognition of the species and subspecies. *Acta Univ. Carol. Biol.*, pp. 37–67.

Power, G. (1978). Fish population structure in Arctic lakes. *J. Fish Res. Bd. Can.*, **35**: 53–39.

Qasim, S.Z. (1973). An appraisal of the studies in maturation and spawning in marine teleost from the Indian waters. *Indian J. Fish.*, **20(1)**: 166–181.

Qasim, S.Z. and Bhatt, S. V. (1966). The growth of thee freshwater murrel *Ophiocephalus punctatus* (Block): *Hydrobiologia.*, **27**: 289–316.

Qasim, S.Z. and Qayyum, A. (1961). Spawning frequency and breeding seasons of some fresh water fishes with special references to those occurring in the plains of North-India. *Indian J. Fish.*, **8**: 24–43.

Qayyum, A. and S.Z. Qasim (1964). Studies on the biology of some fresh water fishes *Ophiocephalus punctatus* (Block). *J. Bombay Nat. Hist. Soc.*, **51(1)**: 74–98.

Quasim, S.Z. (1973). An appraisal of the studies on maturation and spawning in marine teleostes from the Indian water. *Indian J. Fish.*, **20 (1)**: 166–181.

Radtke, R.L. (1984). Scanning electron microscope evidence for yearly growth zones in giant bluefin tuna *Thunnus thynnus*, otoliths from daily increments. *U.S. Fish Wild. Serv. Fish Bull.*, **82**: 434–440.

Rai, B.P. (1967). Cyclical changes in the ovary of *Tor* Barbus *tor* (Hamilton). *Acta. Zoologica*, **48(3)**: 289–308.

Raj, B.S. (1945). The decline of mahseer fishes in Kumaos lakes and a remedy. *Proc. Nat. Inst. Sci. India*, **11**: 341–345.

Rai, U.S. (1947). Effect of weirs at the canal headwaters in the distribution of *Tor putitora* (Ham.) in Punjab. *J. Bombay Nat. Hist. Soc.*, **47(3)**: 449–454.

Rao, V.V. (1969). Sexual dimorphism and colouration among band fishes of the family Copolidae. *J. Bombay Nat. Hist. Soc.*, **66(2)**: 388–390.

Raunsefell, G.A. and W.H. Everhart (1953). *"Fishery Science: Its Methods and Applications."* John Wiley and Sons. Inc., New York, **8**: 382.

Ribble, D.O. and Smith, M.H. (1983). Relative intestine length and feeding ecology of fresh water fishes. *Growth*, **47**: 292–300.

Ricker, W.E. (1946). Production and utilization of fish populations. *Ecol. Mongor.*, **16**: 373–391.

Ricker, W.E. (1971). *"Methods for assessment of fish production in freshwater."* I.B.P. Book, London No. 3, Sec. ed., pp. 1–269.

Royce, W.F. (1963). A morphometric study of yellowfin tuna *Thunnus albacore* (Bonnaterre). US Fish and Wildlife Service. *Fishery Bull.*, **63**: 395–443.

Saether, B.S.; Johnsen, H.K. and Jobling, M. (1996). Seasonal changes in food consumption and growth of arctic charr exposed to either simulated natured or a 12 : 12 LD photoperiod at constent water temp. *J. Fish. Biol.*, **48**: 1113–1122.

Saigal, B.N. (1964). Studies on the fishery and biology of the commercial cat fish of the Ganga river system. II Maturity, spawning and food of *Mystus aor* (Ham.). *Indian J. Fish.*, **11(1)**: 1–44.

Sampath, K. (1984). Preliminary report on the effect of feeding in *Channa striatus. Aquaculture*, **40**: 301–306.

Sandhu, G.S.; Tondon, K.K. and M.S. Johal (1994). Growth studies on an endangered fish *Tor putitora* (Ham.) from Govindsagar (H.P.) India. In: P.V. Dehadrai; P. Das and S.R. Verma (eds.) *Threatened Fishes of India*. Natcon Pub. No. 4: 137-142.

Saunders, L. and G. Power (1970). Sex ratio and fecundity of American smelt, *Osmerus mordax* (Mitchill) from Matamek lake, Quebec. *Canadian J. Zool.*, **48(1)**: 898–899.

Schaefer, M.B. (1948). Morphometric characters and relative growth of yellowfin tuna (*Neothunus macropterus*) from Central America. *Pacific Science*, **2**: 114–120.

Schacfer, M.B. and L. Walford (1950). Biometric comparison between yellowfin tuna (*Neothunnus macropterus*) of Angola end of Pacific Coast of Central America. U.S. Fish and Wildlife Service. *Fishery Bull.*, **51**: 425–442.

Sehgal, K.L. (1971). Fisheries survey of Himachal Pradesh and some adjacent areas with a special reference to trout, mahseer and allied species. *J. Bombay Nat. Hist. Soc.*, **70**: 458–474.

Sehgal. K.L. (1974). Report on the work done by the coldwater fisheries. Research Unit of CIFRI in Kashmir during 1966–71. *Mimeographed*: 1–13.

Sehgal, K.L.(1991). Artificial propagation of Golden mahseer, *Tor putitora* (Ham.) in the Himalaya. *NRCCWF Sp. Publ.*, **2**: 12.

Sehgal, K.L. (1992). Cold water fish and fishery of the Himalaya Part-1. Rivers and streams. *F.A.O. Publication.* p. 150.

Sehgal, K.L. (1994). State of art of endangered, vulnerable and rare cold water fishes of India. In: Dehadrai, P.V.; P .Das and S.R. Verma (eds.). *Threatened Fishes of India.* Natcon Pub. No. 4: 127-136.

Sehgal, K.L.; Kuldeep Kumar and Shyam Sunder (1984). Food preference of brown trout, *Salmo trutta fario.* (Linn.) in Kashmir. *Indian. J. Anim. Sci.* **54(7)**: 676–682.

Sehgal, K.L. and Shyam Sunder (1992). Coldwater fisheries technologies: Estimation of biological productivity of the mountain streams. *NRCCWF Mannual*, **4**: 29.

Sen T.K. and Jayaram, K.C. (1982). Mahseer fishes of India: A review of the large survey of India. *Occ. Pop.*, **39**: 1–34.

Sex W.S. (1957). *Physiology of Fishes.* Academic Press Inc. New York, p. 657.

Shah, K.L. and S.B. Raizada (1977). Preliminary observation on creel census and angling pressure in a stretch of river Beas in Kullu valley. *J. Inland Fish. Soc. India*, **9**: 117–124.

Sharma, M.K. and Rajput, D.B. (1982). Ichthyofauna of Bijnore district (Uttar Pradesh). *J. Bombay Nat. Hist. Soc.*, **83**: 562–569.

Sharma, R.C. (1983a). Qualitative and quantitative seasonal variations in feeding of snow trout *Schizothorax plagiostomus* (Heckel) of Garhwal Himalaya. *Bio. Bulletin af India*, **6(1)**: 20–25.

Sharma, R.C. (1983b). Dynamics of food and feeding habits of *Crossochelius latius latius* (Ham.) in aluvial ecosystem of Garhwal Himalaya. *Comparatively Physiol. & Ecol.* **9(4)**: 305–308

Sharma, R.C. (1986). Food and feeding habit of mahseer, *Tor tor* (Hamilton) of Garhwal Himalaya. *Matsya,* **12–13**: 93–100.

Sharma, R.C. (1987a). Aims and strategies of the fisheries management for the Tehri Dam reservoir of Garhwal Himalaya. *India Reservoir fish Management on Div. in Asia IDRC, Canada,* pp. 45–153.

Sharma, R.C. (1987b). Tropic dynamics of mahseer *Tor putitora* (Teleosten Cyprinidae) in the central Himalayas. *Proc. 5th Congs. Europ. Ichthyol. Stockholm* pp. 145–152.

Sharma, V.K. and Tondon, K.K. (1990). Observation on the morphometric and meristic character of *Cyprinus carpio* (Linn.) from Govindsagar reservoir, Himachal Pradesh. *Pb. Fish Bull.,* **14(1)**: 29–32.

Shetter, D.S. and Hazzard, A.S. (1939). Species composition by age groups and stability of fish population in sections of three Michigan trout stream during the summer of 1937. *Trans. Am. Fish Soc.* **68**: 281–302.

Shobhna, B. and Nair, L. (1974). Observation on the maturation and spawning of *Puntius sarana subanastus* (Val.). *Indian J. Fish.,* **21**: 357–368.

Shreshtha, T.K. (1986). Spawning ecology and the behaviour of mahseer *Tor putitora* (Ham.) in the Himalayan waters of Nepal. In: *Proc. Ist Asian Fisheries Forum"* (ed. Maclean J.L., Dizon L.B. and Hosillor L.V.). Asian fisheries Society, Manilla, pp. 689–692.

Shrestha, T.K. (1988). *Ranching of Mahseer in the Himalayan Waters of Nepal.* Department of Zoology, Tribhuvan University Kathmandu, Nepal, p. 265.

Shrestha, T.K. (1994). Migration of Golden mahseer in Himalayan water of Nepal. *J. Freshwater Biol.,* **6**: 71–77.

Simpson, A.C. (1959). The spawning of the place in the north sea fishery. *Invest. Land. Ser.,* **2**: **607**: 111.

Singh, H.R.; Badola, S.P. and A.K. Dobariyal (1983). Geographical distributional list of Ichthyofauna of the Garhwal Himalaya with some new records. *J. Bombay Nat. Hist. Soc.*, **84(1)**: 126–132.

Singh, H.R. and Dobariyal, A.K. (1983). Morphometric characters and their relationship in the catfish *Pseudencheneis sulcatus* (McClleland). *Indian J. Anim. Sci.*, **53(5)**: 541–546.

Singh, H.R.; Dobariyal, A.K. and Nautiyal, P. (1985). Spawning patterns and environmental regulation of spawning in hillstream fishes. In: *The Endocrine system and the Environment*. Ed. Follett. B.K. *et al.*, Japan Sci. Soc. Press. Tokyo. Springer-Verlag, Berlin, pp. 1–11.

Singh, P.R. and H.R. Singh (2000). Morphometric relationships of *Labeo calbasu* (Ham.) from the river Ganga at Allahabad. *Him. J. Env. Zool.*, **14**: 93–98.

Singh, R. (1966). Morpho-histological studies in alimentary canal of *Bagarius bagarius. Agra Univ. Ras. (Sci.)* **14(3)**: 69–77.

Sinha, Rajesh Kumar; M.D. Hadique and Dilip Kumar Singh (1999). Length-weight relationship and condition factor of a thermal spring loach, *Lepidocephalichthyes guntea* (Ham.). *Him. J. Env. Zool.*, **13**: 131–134.

Smith, F.W. (1947). Sex of mahseer and Bokar caught in Northeast India. *J. Bombay Nat. Hist. Soc.*, **46(4)**: 734.

Smith, I.P.; Matcalfe, N.B.; Huntinford, F.A. and Kadri, S. (1993). Daily and seasonal patterns in the feeding behaviour of Atlantic Salmon (*Salmo salar* L.) in sea cage. *Aquaculture,* **117**: 165–178.

Smith, O.R. (1947). Returns from a natural spawning of cut throat trout and eastern brook trout. *Trans. Am. Fish. Soc.*, **74**: 281–296.

Solomon, D.J. (1977). A review of chemical communication in fresh water fish. *J. Fish. Biol.*, **11**: 369–276.

Srivastava *et al.* (1985). *Inland fish marketing in India: Overview summery and conclusion*. Concept Publishing Co., New Delhi, p.151.

Spanishlab, L. (1968). Sexual maturity, Sex ratio and fecundity in brown trout, *Salmo trutta M fario* L. in the Lucka river. *Zool. List,* **17(3)**: 253–258.

Steven G.A. (1930). Bottom fauna and the food of fishes. *J. Mar. Bio. Assoc.*, London **16**: 677–700.

Sundar Shyam and C.B. Joshi (1977). Preliminary observation on the spawning of *Tor putitora* (Hamilton) in Anjii stream, Jammu Province during 1969. *Inland J. Fish.*, **24(1&2)**: 153–158.

Swamp, K. and A. Swaroop (1975a). Sexual dimorphism in the catfish *Mystus vittatus* (Block). *Curr. Sci.*, **44(1)**: 16–17.

Swamp, K. and A. Swaroop (1975b). Sexual dimorphism in *Pseudeutropius artherenoides* (Block). *Curr. Sci.*, **44(16)**: 588.

Swamp, K. and Singh, D.K. (1971). Sexual dimorphism in the buffer fish *Chelondon cutcutia. Curr. Sci.*, **40**: 39–40.

Swamp, K.; S. Srivastava and V.K. Das (1972). Sexual dimorphism in spring eel, *Mastacembalus puncalus. Curr. Sci.*, **41(2)**: 68–69.

Swarup, K. (1968). Sexual dimorphism on the gar fish *Xentodon cancilla. Curr. Sc.*, **37(1)**: 352.

Swarup, K. and S. Srivastava (1972). Sexual dimorphism in the *Giant gaurami* and *Colisa fasciatus* (Anabantidae). *Proc. Nat. Acad. Sci. India*, **42b**: 61–62.

Talwar, P.K. and Jhingran, A.G. (1991). *Inland Fishes.* Vol. 1. Oxford and IBH publishing Co. Pvt. Ltd. New Delhi, p. 541.

Thakre, V.Y. and Bapat, S.S. (1981). Maturation and spawning of *Rasbora daniconius* (Ham.). *J. Bombay Nat. Hist. Soc.*, **72**: 38–45.

Thomas, H.S. (1873). *The Rod of India* C. Stoiz Mangalore, pp. 3-9.

Thomas, H.S. (1893). *The Rod in India.* W. Thacker & Co. London, p. 692.

Thomas, H.S. (1897). *The Rod in India* 3rd ed. W. Thacker and Co. London, pp. 435.

Tobar, J.G. (1974). A contribution to the study of *Lates nelotics* Nile perch in lake Chadh. Federal Ministry of Information Printing Division Lagos. *Occassional Pape.*, **17**: 1–23.

Tondon, K.K. and M.S. Johal (1983). Study on age and growth of *Tor putitora* (Hamilton) as evidence by scale. *Indian J. Fish.*, **30**: 171–175.

Tondon, K.K. and M.S. Johal (1996). *Age and Growth in Indian Freshwater Fishes.* Narendra Publishing House, p. 320.

Tondon, K.K.; M.S. Johal and Kaur, J. (1983). On the systematic age and growth of *Labeo dero* from Govindsagar, Himachal Pradesh. *India Vest. cs. Spolec Zool.*, **53**: 54–65.

Tondon, K.K., M. S. Johal and Kaur, J. (1993). Growth parameters of an endangered species Golden mahseer, *Tor putitora* (Hamilton) from Govindsagar. In: *Advances in Limnology* (H.R. Singh ed.) Narendra Publishing House, New Delhi, pp. 207–220.

Toor, H.S. (1964). Biology and fishery of *Lenthrinus lentjan* Lacepede Pt. I-II. *Indian J. Fish*, **11(2)**: 581–685.

Valsangkar (1993). Mahseer fishery of Koyna river (Shivaji Sagar) in Maharashtra: Scrap to Bonanza. *Fishing Chimes.*, **12(10)**: 15–20.

Van Oosten (1929). Life history of like hearing *Leuceechthys article* of Lake hour as revealed by its scale with a critique of the scales method. *Bull. U.S. Bur. Fish.*, **44**: 265–428.

Vasudevappa, C. and James, P.S.B.R. (1980). Maturity and spawning of marine catfish, *Tachsurus dussumeiri* (Val.) along the south Canara coastal. *Proc. Indian Natn. Sci. Acad.*, **B46**: 90–95.

Vijay Soni, (1992). A brief note on the mahseer conservation plan 1991–further details on discussion. *Pub. Fish Bull.*, **16(1)**: 71–72.

Vladykov, V.D. (1934). Environmental and taxonomic characters of fishes. *Trans R. Com. Inst.*, **20**: 99–140.

Walker, W. (1888). *Angling in Kumaon Lakes*, Calcutta, p. 39.

Ward, H.B. and Whipple, G.C. (1959). *Freshwater Biology*. Edmondson (Ed.), New York, p.1448.

Ward, J.V.; S.P. Canton and L.J. Gray (1978). The stream environment and macroinvertebrate communities, Contrasting effect of mining in Colorado and the eastern United States. In: J.H. Throp and Giddon, J.W.(eds.) *Energy and Environment Stress in Aquatic Systems*, Doe symposium series (comp. 77114) spring fields, pp. 176–187.

Welch, P.S. (1948). *Limnological Methods*. The Blankistan to Philadelphia, pp. 1–381.

Wiencaszed, B. (1992). Biometric characteristic of Baltic eel-pout *Zoarces viviparus* (L.) from the Pomeranian Bay. *Acta. Ichthyol. et. Pisc.*, **22** Fasc: 39–57.

Wood, H. (1930). Scottish harring shoals pre spawning and spawning movement. *Scotland J. Fish.*, **3(1)**: 59–90.

Index